A FINE SMIRR OF RAIN

*Smirr: A Scottish variant of smurr,
fine rain or drizzle.*

ALSO BY WILLIAM BRIDGES

Poetry

Common Places

Weedpatch or Jericho?

The Arafura Sea

The Perfect Country of Words

Eye

*The Landscape Deeper In:
Selected Poems, 1974-2004*

Other

*Dear Viola: Reporting, Writing and Editing
for the Student Journalist*

*Under the Heaven Tree:
An Indiana Childhood*

Five-Mountain Morning: A Memoir

"A Fine Smirr of Rain," by William Bridges. ISBN 1-58939-941-2.

Published 2006 by Virtualbookworm.com Publishing Inc., P.O. Box 9949, College Station, TX 77842, US. ©2006, William Bridges. All rights reserved. No part of this publication may be reproduced, stored in a retrieval system, or transmitted in any form or by any means, electronic, mechanical, recording or otherwise, without the prior written permission of William Bridges.

Manufactured in the United States of America.

A Fine Smirr of Rain

Variations on a Theme

WILLIAM BRIDGES

"In from the West a fine smirr
Of rain drifts across the hedge.
I am only out here to walk or

Make this poem up. . . ."

— W.S. Graham
"What is the Language Using Us For?"

PHOTOGRAPHS BY NANCY L. BRIDGES

COVER BY LINDSAY HADLEY AND TIM LISKO

FOR SUSANNA

CONTENTS

1

ANCIENT RAIN

ON CHUNGSHIAO ROAD in Taipei there is (or was a few years ago) a tiny teahouse. The façade is no wider than its glass door, and a barrel awning shelters four feet of sidewalk. A paper lantern hangs under the awning, on the end of which, above the curb, are four archaic Chinese characters, the teahouse's only identification.

I couldn't decipher the writing, nor could the more fluent speakers to whom I pointed it out. I finally copied the ideograms and took them to a scholar. "They're about very old rain," he said. "Ur rain. Rain from the beginning of the universe." A simple translation, he said, would be the Ancient Rain Café. It sounded like the gateway to a parallel universe.

But the café did not then vanish, as I half expected it to. Later a friend and I went inside, walking down a narrow hallway, leaving our shoes on a shelf, and entering rooms that expanded behind the neighboring businesses. The walls were hung with calligraphic scrolls. We sat on cushions beside low tables where very good green tea was served in earthenware pots. It was a cozy place, safe from whatever rain—real or metaphoric—might be beating down outside. I never learned how it got its name, but the words "ancient rain" have echoed in my mind ever since.

When *did* it start raining? The question is imaginative, not scientific. There is a science of paleometeorology, but it deals more with dust than rain. Plug it into the internet and you get "ancient windblown sandstones" and "eolian dust deposition in central Asia." Rain is an inexact term, but in any usual sense it's always been there. It comes close to being a universal experience, missed only by those in the most extreme climates, those who die very young, and occasional ombrophobes, who still must have seen enough of it to be frightened. The Roman emperor Carus died ambiguously during a rainstorm; he was either struck by lightning or his chamberlains set fire to his tent to make it look that way. Three million years ago, in the Afar Depression of East Africa, Lucy looked out and decided to spend the day under cover. Animals also experience rain. Birds can fly in it, but don't like to (except for a few like the Manx shearwater). On the Hebridean isle of St. Kilda, Soay sheep—smarter than the average—take refuge in stone sheds called cleits. Cows and horses often stand patiently in the rain as did, probably, woolly mammoths and dinosaurs. (However, reindeer don't like it and get foot rot if they stand in boggy pastures.)

So when was the first rainstorm? "I think it was on the eighth day," a Scottish friend writes. "Then God looked out the window and said . . . ." But Genesis has Him encountering water on the first day, and this looks more and more like what happened. Find when water began and you'll know when it began raining, simple as that.

Earth's first epoch, the Hadean or Hellish, was once thought to be an eternity of waterless, heaving magma. But now it seems things may have cooled down more quickly; water and oceans — and rain — may have been around earlier than anyone thought. The clues to this are some very old zircons from the Jack Hills of Western Australia. In 1999, researchers led by Dr. Simon Wilde identified zircon grain W74/2-36. Forget necklaces and brooches; although an attractive deep purple, W74/2-36 is less than a hundredth of an inch across, the width of a couple of hairs. Its finders dated it by measuring how much of its uranium had decayed into lead. It turned out to be 4.404 billion years old, give or take eight million years. Scientists prefer the terms "solid" and "mineral," but from a lay perspective it is the world's oldest rock. Zircons, not diamonds, are forever.

Zircon W74/2-36 shook things up a bit, and not just by its age. It carried a ghostly message from the Earth's first crust: "cooler than you thought, cool enough for water and land." *Nature* magazine, on its cover, called W74/2-36 "A chip off the old continent." The message was delivered in a somewhat meandering way. Almost all zircons formed in granite that evolved when oceanic crust slid downward under a continent. W74/2-36 began life near the sea and then rode along as its granite cooled again, reared into mountains, weathered away, and vanished, leaving its zircons in the sands of an ancient Australian stream bed. "These zircons have really been through the wringer," researcher William Peck told a NASA reporter.

Zircon W74/2-36 and its slightly younger siblings are also rich in "heavy oxygen," another marker for contact between rock and water at relatively low temperatures. All this raises the possibility of very early life, although one writer adds quickly, "We need more grains, and lots of them."

How ancient is rain? Scientists have dated fossil imprints of raindrops in India to 1.6 billion years ago. Surface water may have been around for 4.4 billion years, only 150 million years from the beginning, and rain is probably just as old. The atmosphere once held as much water as the oceans do now. The earliest torrential rains would have turned to steam upon hitting the still-hot earth. This steam would have gone straight back up to release heat into space, like a celestial air-conditioner. Hot seas would have collected and then perhaps been vaporized by comet and meteorite strikes, along with any life they held.

Early rain was not good for frolicking—no April showers in the Hadean. Besides the water that comets and meteorites brought with them, volcanoes were exhaling vapor mixed with nitric, sulfuric, and hydrochloric acid, methane, ammonia, carbon monoxide, and "minor other nasties," a geologist friend says. The air contained practically no oxygen, compared to today's comfortable 16 percent. If you're depressed by a cloudy day, you would have been extremely unhappy (as well as breathless and dead) in the Hadean. It was raining like hell—"several hundred million years of falling rain" to form the oceans, says one source. Whether the downpour was unbroken, only God knows and He wasn't talking until later.

In October, 2005, I checked back with Dr. Simon Wilde, the zircon man, who replied that "what we are now prepared to say is that there was probably surface water early on at 4.4 to 4.36 billion years—but that it was clearly widespread by 4.2 billion years, when many zircons show elevated oxygen levels. We have recently suggested that 4.2 billion years should mark the boundary between the Hadean and Archean, since this is the time when continental crust was being 'seriously' recycled for the first time. It's all good fun."

In a strange way, the toxic rain of this early period was beneficent. A pair of Japanese scientists, Takafumi Matsui and Yutake Abe, have demonstrated how little difference there was in the beginning between the atmospheres of Earth and Venus. Somewhat different vapor pressures allowed the release of rain onto the earth, but turned Venus into a super-heated hell.

It has been raining on Earth for a very long time. One internet article even refers to "ancient rain," but this turns out to be the descent of tiny particles—oxidized from oceanic iron—to the primordial seafloor. Today these particles show up as hematite bands in rock. We live not only on a wet planet, but on a rusty one.

* * *

Rain fell early in the afternoon, but by the time I left the house it had largely run off, dried, or collected in puddles or on the small branches of trees and bushes. There seemed an optimum size: the smallest trees, some hardly more than shoulder height with a fretwork of tiny branches, were the most jeweled. Some drops still clung to the larger linden tree in the front yard, at the points where downward looping twigs turned up again, or around leaf buds. Looking closely through the larger drops, I could see the house and trees across the street, inverted. Down the street, a child leaned over a porch rail, her palm up. It was the first day of spring.

2

THE SHAPE OF A RAINDROP

MEASURING A RAINDROP is like harnessing a flea. It can be done. I once saw, in a Middle European town, a flea circus in a wooden box. The fleas had little harnesses with which they pulled minute wagons and made wheels go round. I was glad I had seen them, but I didn't go back for the evening performance.

When it comes to raindrops, most people don't see what hits them. For all their ubiquity, raindrops are hard to distinguish as they fall. In a storm, one sees a lot of flying water, but no discrete drops until they hit something. In a light shower, you can hold out your hand and *feel* raindrops, but you won't see anything. It's a slightly eerie sensation, like having words spelled into your palm by a ghost.

The Roman poet Lucretius promised to explain rain but only speculated. Modern scientists know a lot about it, including the size, shape, and velocity of raindrops, but this knowledge is recent. "Cloud science" developed in the 1940s, and the first significant article on raindrop shape seems to have been by Dr. Athelstan Spilhaus in 1948. In the same year, two scientists devised the Mitchell-Palmer equation, relating raindrop size and distribution to the intensity of the downpour.

Fifty years earlier, though, two researchers—one at a German university, the other on a Vermont farm—compiled some data. Using a vertical wind tunnel, Dr. Philipp Lenard (later a favorite scientist of Hitler's) got water drops to float before his eyes. The Vermont farmer, Wilson A. Bentley, found that if he stood outside with a pan full of flour, raindrops would form drop-sized dough balls in the bottom. He could let them dry and measure them on his kitchen table. He checked the method's accuracy by releasing water droplets of a known diameter from stepladders and the roof of his house. You can try his experiment at home, perhaps when no one is looking.

Bentley is best known for taking 5,381 photomicrographs of snow crystals between 1885 and 1931. He was called Snowflake Bentley, and his neighbors thought he was a little odd. But he may have been the world's most heroic amateur weather observer. Besides snow, he investigated clouds, rain, hail, dew, frost, and rocks. During 49 years, he recorded 634 auroras. With a fine thread, he tied a grasshopper to a twig so he could photograph the dew that collected on it. Then he turned the creature loose, perhaps to wonder, as a biographer suggests, "about what strange thing had happened to him the night before."

I am trying not to be overinformative.

Raindrops range in diameter from a fiftieth to a quarter of an inch. Less than a fiftieth and it's drizzle. My favorite unpublished novel opens with the words "Drizzle becomes rain," and the author is exactly right.

Up to about a twenty-fifth of an inch, raindrops are spherical, because surface tension keeps them that way. But above this size, we face the teardrop vs. hamburger bun quandary. The first to use the bun simile may have been James E. McDonald (please, don't say it), writing in the *Scientific American* of February, 1954. He also complained that cartoonists and commercial artists — to which one might add weather forecasters — portray raindrops as tear-shaped. They are not. Larger raindrops are flattened on the bottom and curved on top. Not even tears are tear-shaped. A professor of meteorology has a "Bad Rain" web site with a circle-and-bar symbol superimposed on the conventional teardrop. To those who say the teardrop "communicates" better, he replies that bad scientific communication is worse than none. Still, after 50 years of repeated use, the hamburger bun may be a soggy simile. Maybe it's time to find another—one that doesn't evoke the image of a skyful of tiny Big Macs.

Connoisseurs of rain can get into it (wade into it?) as deeply as they want. An article in the *Journal of the Atmospheric Sciences* runs for 18 pages and is titled "The Microphysical Structure of Extreme Precipitation as Inferred from Ground-Based Raindrop Spectra." McDonald's 1954 article, "The Shape of Raindrops," was shorter—a sparkling essay on the aerodynamic forces that mold a raindrop into a bun. McDonald demonstrated that these forces affect a raindrop much as they do an aircraft and are in continual flux to keep the drop stable.

"It seems almost unfair," he wrote, "that the fate of so carefully equilibrated a little system may be no more glorious than to splatter down on some dusty road at the beginning of an August thunderstorm."

McDonald wrote "at the beginning" advisedly, since the first drops in a thunderstorm are the largest and travel the fastest, up to 30 feet a second. This "terminal velocity" is constrained by the atmosphere. Otherwise, a raindrop falling from six miles up would be traveling 300 m.p.h. and anyone stepping out would soon be as full of holes as Frankie Gusenberg after the St. Valentine's Day Massacre.

THE SHAPE OF A RAINDROP

Any good encyclopedia will tell you how vapor molecules are drawn to a microscopic nucleus of sea salt, dust, or even bacteria. (In northern latitudes, it's often ice; much rain is melted snow.) The clumps become tiny droplets that fly around in the cloud, running into each other by the billions, until enough of them stick together to form raindrops, and the raindrops get big enough to fall. Drops take from a few seconds to 15 minutes to reach the ground. On the way down, they smash into other drops and get bigger. But much bigger than a quarter of an inch, and they break up.

The amounts of water are vast. Each year, about 140,000 cubic miles of water fall on the Earth, three-fourths of it over oceans. This is four billion gallons a second, but since roughly the same amount evaporates every second it all works out nicely.

There remains the perennial question of whether one stays drier by running through the rain or by walking. Does the runner's thrust-out chest encounter more raindrops than the stooped form of the stroller? You should not be surprised that this has been studied, and that the answer is, "Run, don't walk." But the Zen Buddhist answer is somehow more satisfying. When Richard Baker succeeded Shunryu Suzuki as leader of the first American Soto Zen community, he told him:

A FINE SMIRR OF RAIN

Walking with you in Buddha's gentle rain
our robes are soaked through,
but on the lotus leaves
not a drop remains.

* * *

Driving to the library in the rain to get a book about rain, I passed an involuntary field researcher. This was a youth of 18 or so who was soldiering along through the downpour with no hat or umbrella. His T-shirt and black polyester windbreaker were soaked. He looked a little like James Dean, ready to rebel. Later, in the supermarket, a cheerful stock girl said, "Wet out?" At the post office, I looked up into the rain and noticed birds nesting behind the final "E" of UNITED STATES. Coming up the driveway at home, I saw that the willow at the bottom of the garden had become, overnight, a green mist.

3

Remembering Clouds

ON THE EVENING of May 12, 1994, coming home from the American Library in Taipei, I saw the planet Jupiter sailing above a rack of mauve clouds. This was during a break in the city's perpetual haze of pollution. I almost never saw stars or planets in Taipei, but there were plenty of clouds, except when the front preceding a typhoon had swept them away.

I remember other clouds: great summer towers rising above our garden on Utica Pike in Jeffersonville, Indiana; the animal shapes my mother pointed out during a drive to Niles, Michigan, when I was five; cloudfields over Peru, through which we looked down, hoping for a glimpse of the Amazon. From a boat off the coast of Cornwall one summer day, the white, fair-weather clouds looked unusually soft and melting. I don't know if this is characteristic of Cornish clouds or per-haps of seaside clouds generally. I took pictures of them — too many pictures. Throwing some away later, I cut out one of the clouds and pasted it in a writing notebook, where I found it again after a search, and the date—Thursday morning, Aug. 14, 1997, on our way to the Isles of Scilly. I still have pictures of a dazzling column of moonlight falling from a cloud to the sea off the island of St. Kilda in the Outer Hebrides.

Cloud memories are common. My father, flying to Hawaii after his second wife's death, saw "the clouds, like snowfields spreading in all directions . . . . It was only when we were well down that the ocean and the island were visible in the failing light." He recorded these facts in a travel diary and dated them: December 16, 1970. He was five years younger than I am now.

Annie Dillard has written about such observed and dated clouds. In her meditation on transience and eternity, *For the Time Being*, she describes poignantly the clouds painted by John Constable in 1828 as his wife lay dying. "Maria Constable died that November," she writes. "We still have these dated clouds."

Clouds have been observed and commented on throughout human history; Babylonian weather forecasts based on them have been preserved from the 7th Century B.C. In 349 B.C., Aristophanes took third prize in an Athenian competition for his comedy, *The Clouds*. This drama shows that the ancients knew quite well how clouds draw up water from the

Earth, even if they lacked scientific explanations. Strepsiades, a would-be Socratic disciple, is surprised when the clouds appear on stage as a chorus of goddesses; he had thought they were "only fog, dew, and vapor." Also, he objects, these women look nothing like clouds—they even have noses. But Socrates reminds him that clouds can take any form—foxes, bears, even women. (Modern productions of the play sometimes feature women in white gowns with cotton wool on their heads.)

The study of clouds didn't become much more scientific than this until 1802, when a young English pharmacist, Luke Howard, gave them Latin names—cumulus, stratus, cirrus, nimbus. With a few additions, his list is still used today. Scientists have now studied clouds exhaustively and know a great deal about them, some of which I will get to later.

For most people, clouds are a backdrop, barely noted unless they become spectacular. But England has a Cloud Appreciation Society, with a stylish website, that vows war on "blue-sky thinking." Its founder, Gavin Praetor-Pinney, encourages correspondence, "but only if it is about clouds. Otherwise we're not interested."

Others take a less light-hearted approach. There is a Working Group on Layered Phenomena in the Mesopause Region, which studies, among other things, noctilucent clouds.

These are exceptionally high cirrus clouds, 75 to 100 kilometers up, that glow occasionally in the night sky. They are made of ice crystals or, some think, volcanic dust.

Still more serious are people afflicted with nephophobia, the fear of clouds. A website devoted to therapy for them promises relief and points out, cogently, that any object associated with a supremely unpleasant experience can trigger a phobia. One imagines childhood terror at a thunderstorm, or an irrevocable disaster on a cloudy day.

Luke Howard, who was a Victorian list-maker before the Victorian era, perpetrated a bit of a fraud when he named the clouds. Clouds are not as discrete as his four-tier system might suggest. Like many other classifiers, he imposed labels on a continuum, and the labels soon began to define reality. Something about the human mind prefers neat lists and divisions to the messiness that's really out there. The spectrum has hundreds of named colors but countless others in between. Edmund Burke, the English statesman, said he could tell day from night, but was never sure when twilight began or ended.

Still, if you want to know about clouds, it helps to have some labels, and Howard's four categories are a start. Cumulus, stratus, nimbus, cirrus. Heaps, layers, rainmakers, curls. The last simplified labels are from *The Book of Clouds* by John A. Day, a Minnesota octogenarian known worldwide as the

Cloudman. His heaps are fair-weather clouds, unless they become thunderheads and turn black at the bottom. "Black walls," my wife calls these, as she heads for the basement (by which time our weather-alert radio is squawking every five minutes).

The layers are less defined. These are pale sheets and riffles of cloud on their way to becoming overcast; they mask the sun and moon and produce mist and ground fog. Rainmakers (nimbus) can be part of the dramatic and dangerous thunderheads. But mixed with layers they're the least appealing of clouds. It's going to rain all day, dully, miserably. Forget the picnic. Curls (cirrus) are technically rainmakers, but they're so high that the rain usually becomes *virga*, evaporating before it hits the ground. These are the mare's tails that frisk around the sky on a sunny day.

You can predict the weather, a little, with no more information than this. Ordinary wisps and heaps, mare's tails, thin layers—your golf game is probably safe. Thickening layers, halos, and masking of the sun—take your umbrella. Whittier got it right in "Snowbound": "The sun that brief December day/Rose cheerless over hills of gray,/And, darkly circled, gave at noon/A sadder light than waning moon." He was probably looking at altostratocumulus.

Day's book is also helpful for odd clouds and atmospheric effects—glories (giant shadows thrown against the clouds), sun dogs, crepuscular rays, auroras, and lenticular or flying-saucer clouds. And rainbows. From a train in Scotland on Aug. 25, 2005, I saw a perfect 180-degree rainbow, with both ends in clear view, bathing fields and farmhouses in a golden haze.

After spending an hour one morning boning up on Day's system, I walked to the post office at 10 a.m. under a blue sky with only a wisp or two of cumulus. At the horizon the blue turned slightly dirty. (In Taipei, even on a fair day, the air at the end of the street was often brown.) By noon, fair-weather cumulus clouds covered the sky. By suppertime, these were gone, replaced by dimpled sheets of altocumulus, cirrus, and layer clouds across the setting sun. When I came back in the house, I told my wife that it looked like rain the next day. "You're right on it," she said, pointing to the TV where an orange and green arc was marching toward us across Iowa.

Coming down through clouds once, a plane I was on coasted for half an hour through pastel canyons, so close that a wing-walker might have stepped out into them. In the early days of rain research, scientists did fly Piper Cubs into them, leaning out the window to capture raindrops on coated plates. Only a few years later, astronauts gave us the first pictures from space of the entire Earth, with great whorls of cloud drifting across a blue and brown planet, all that we have.

Looking at those photos, it's clear Socrates was half right when he told Strepsiades that the clouds were moved by "the aerial whirlwind," not the gods, and that rain came when clouds bumped together and burst. Strepsiades was dumbfounded by this revelation of the New Science. "I always thought it was Zeus, pissing into a sieve," he said.

* * *

Near midnight, I went out to see a full moon sailing in a sky of nighttime cumulus, an Albert Pinkham Ryder painting. Its rim seemed a darker gold, the raised edge of a coin wreathing the face of an emperor. The light dimmed and disappeared as dark clouds slipped over the disc. When it emerged again, the brightness began as an illumination, a boiling of light, at the cloud's trailing edge. Once, high up, a single star appeared.

4

THE SOUND OF RAIN

IF A RAINDROP FALLS in the middle of the Pacific Ocean, does it make a sound? This is always a matter of definition. But there is something infinitely lonesome about the idea of rain falling with no one to hear it.

On an old farm, those living in the farmhouse hear rain hitting the roof or windows. Behind the house is a barn or two, maybe a chicken coop, in which animals drowse to the beat of rain. Back of these are sheds that were put up once for some purpose, but now are abandoned and falling down, crypts for broken machinery. And behind the sheds, where no one goes, the rain still falls, making or not making a sound. Almost all rain falls without being heard, at least by humans.

Rain by itself makes no sound that we can hear. It is the drummer's stick, percussing whatever it touches, the taut drumhead or the edge of the cylinder, a rim shot, *clickety click*. Australian aborigines have rainsticks—crooked sticks three or four feet long whose pith, part of it, has been eaten away by termites. Now there are many loose fragments inside the stick, and when the aborigine picks it up there is the sound of a little shower of rain that can go on for several seconds *shshshshshshshhhhh . . . tik . . . . . tik.* An Australian dealer offers rainsticks made, he says, by burying the sticks in the ground on a mountain during a storm and waiting for lightning to produce "sticky rain crystals" within them, all the while calling upon gods and goddesses. You may believe this if you wish. The newspaper I edited ran weather reports from its staff meteorologist, Raynor Schine; I cannot throw rocks at other fantasists.

As rain diminishes, you wait for the last raindrop off the eave . . . . *tik.* Or a more hollow sound, depending. I once lived for a week in a very cheap hotel room in Hualien, Taiwan. This was on the east side of the island, which gets the typhoon winds off the Pacific. It didn't rain a lot while I was there, but when it did my bare little room—a cell, really—resonated like the sound chamber of a violin, or more accurately like one of those stringed instruments made from a cigar box or a coconut . . . *tok . . tok . . tok . . . . tok.* I could feel my mind slow down and stop. I was trying to write, but some days I didn't get much done.

THE SOUND OF RAIN

That rain seemed quiet, but the total volume of noise in a large storm is incalculable, even aside from lightning hitting things, and thunder. Some weather writers describe the power of storms through comparisons: "Several times the energy of the atomic bombs dropped on Japan," they say, as if we would understand how much that was. On my newspaper, we used to joke blackly about this cliché: "The train crash (or propane-tank explosion or building collapse) 'sounded like an atomic bomb,' said Hideki Minoru, a former resident of Hiroshima." If you are inside something like a tent or a corrugated silo, the sound of weather is magnified and can be terrifying. Burrowed in our tent, my family and I once waited out a mountain hailstorm in Colorado, wondering if we were about to be shredded.

If you can't get outdoors, you can call up a web site for drizzle, which sounds like bacon frying over a low fire. Or you can order some rain on a relaxation tape or CD. During a dry spell, I sent off for a thunderstorm. The cats didn't know what to make of it, but came into the room instead of hiding in the closet as they do in real storms. After a while, I went to sleep. But before that I analyzed the sound, or at least the way the microphone recorded it. In the distant background was a steady, undifferentiated rustle, like radio static. Nearer to me were layers with more discrete sounds, but still mostly uniform. In the sonic foreground were soft sizzles, pops, and snaps as the raindrops splattered on nearby objects. After a few minutes the sound of running water began. After a few more minutes, my wife said, "You know, if someone was incontinent, there could be a real problem here."

About all that unheard ocean rain. If a raindrop falls in the Pacific, someone *may* be listening. Since the 1990s, scientists have used acoustical rain gauges (or ARGs) to record rain, wind, and salinity in the uttermost parts of the sea. An ARG is moored somewhere — perhaps in the South China Sea — and an onboard computer wakes it up now and then to listen for rain. Since 1997, a satellite of the U.S.-Japanese Tropical Rainfall Measuring Mission has also been scanning for rain and other phenomena. Some tropical rain, like that from cirrus clouds, evaporates as it falls and becomes virga.

One can measure oceanic rain by its sound because each drop has a two-part signature — the splat when it hits and the "ringing" air bubble it produces underwater. An electronic ear can distinguish these sounds from waves, man-made noises, and fish, although colonies of snapping shrimp are a problem. An underwater bubble begins life as a "screaming infant," but soon becomes a "quiet adult." The newborn bubble oscillates wildly as it tries to equalize its internal pressure with that of the water, producing a ringing high-frequency tone — the scream. Small bubbles ring at a higher frequency than big ones. Research into ringing bubbles goes back at least to 1933 when a certain M. Minnaert published an article on "Musical Air Bubbles and the Sound of Running Water," in the *London, Edinburgh, and Dublin Philosophical Magazine and Journal of Science.* Minneart is still being cited in acoustic and oceanographic literature—he appears to have been the noted Utrecht astronomer Marcel Minnaert, tossing off a little classic on sound in between work on stars.

Listening to raindrops ring in mid-ocean may seem an esoteric pastime, but it provides a clearer picture of global weather. The tropics account for two-thirds of the rain that falls, much of it over oceans. The formation of a raindrop releases latent heat that is a main driver of atmospheric circulation.

Rain also makes wonderfully unscientific sounds. Two thousand years ago, the Dongson people of Southeast Asia were making bronze rain drums. A Thai general is said to have placed one under a waterfall to mimic the sound of a marching army. Some of these drums are three feet tall and look like charcoal braziers. They have a thin bronze tympanum on which a drummer can play. Or you can just leave them out in the rain. They are an example of what an old encyclopedia calls "instruments of indefinite sonorousness," like the rain itself. I saw a reference to "raindrop drums" in promotional material for a Jetset recording of "Macha Loved Bedhead." A Jetset publicist told me the idea of a drum played by raindrops was poetic license, but added, "If the drummer for Macha could work out a way to do that, he would."

* * *

A cold Easter morning, the first day of my 70th year, the sky an unrelieved gray. There was not supposed to be rain, but a few drops fell anyway, spotting sidewalks and the steps of churches. A trace, meteorologists call it, unmeasurable. Instead of rain, we had a blizzard of Bradford pear blossoms along all the streets. In the afternoon, I built a hearth fire, to warm the last of winter on its way.

5

DOES GOD SMELL LIKE RAIN?

OF COURSE RAIN HAS A SMELL. We all know that—and like other things we know, it turns out not to be so simple after all. Most of us are just thinking of the freshness of the Earth after a storm, what a poet calls "the rain-sweetened air." Or it is the indefinable something, intuition as much as odor, that tells us rain is on the way. I find only a few references to smell during the storm itself, which may tell us something, perhaps that raindrops themselves are pretty much scentless.

Rain had been threatening vaguely all day, although the early clouds were too thin, the overcast too light, actually to produce it. ("It can't rain, the clouds are too close together," my father used to say.) So when I went out sniffing about 8 a.m., I didn't detect much except a slight vegetable odor that was probably from the dew still on the grass. The dew itself, when I dragged my hand through it, had no smell. Later, I walked past a house with a damp front yard, but the smell was more of mud than of rain. And aren't these smells all mixed together anyway?

Not quite, it seems. One after-the-rain smell — strongest in woods — is produced by actinomycetes, soil bacteria whose spores are kicked up by rain like an aerosol and strike our noses as a pleasant, fresh smell. A similar smell, more common in dry climates, arises when rain dissolves plant oil deposited by the air on rocks and soil. (Actinomycetes, I am informed, also account for the distinctive smell of the Glasgow subway.)

In 1964, two Australians, I.J. Bear and R.G. Thomas, studied this plant oil and named it *petrichor*, after the Greek *petros*, rock, and *ichor*, the substance the Greeks thought coursed through the veins of the gods. Petrichor has had a limited vogue among word mavens, who tend to apply it indiscriminately to all rain smells. An English blogger named Beth writes winningly on greenfairy.com:

> *This is amazing. While we were driving home today, we drove into a rain cloud — saw it before we hit it. It was smudgy grey. The rain hit the windscreen hard once we were in the middle of it — we were enveloped in petrichor. (At the time I said, "That's a lovely smell," but now I know how to be a little more specific.)*

DOES GOD SMELL LIKE RAIN?

Bear and Thomas, writing in *Nature*, were more precise and identified petrichor as a complex chemical deposited by nearby vegetation on rocks and soil, and released by the fall of rain. They exposed rocks to a controlled atmosphere for as long as a year and then steam-distilled the petrichor, a yellowish oil later found to have 50 or more chemical components. (Is anyone else as captivated by the purity of this science as I am?)

Until their work, scientists had referred to the "argillaceous [clayish] odor" exhaled by wetted clay. Bear and Thomas found that other minerals than clay had this property and coined "petrichor" to set matters straight. They also reported that a perfumery in Kannauj, India, 80 miles west of Lucknow, had for years been extracting the scent from baked clay and selling it as *matti ka attar*, or earth perfume. A year after their 1964 paper in *Nature*, Bear and Thomas wrote that petrichor appears to retard the germination of seeds planted just after a rain. Then they seem to have dropped the subject.

Petrichor has even made its way into literary criticism, with a reviewer writing of author Anita Agnihotri: "Her prose breaks into passages of lyrical beauty that come as a sorely needed revivifying petrichor amid the pitiless glare of callousness and cruelty."

An internet diarist named Petrichor has a blog site titled "Terribly Sane." The internet, in fact, is awash in rain and its smells. Googling "smell of rain" turns up nearly 10 million references, and narrowing the topic sharply still produces thousands. After a discussion on one weblog about bottling the smell (apparently the Italians as well as the Indians have done this), MikeE remarks, "I'd be worried about overdosing — what with this and the nutmeg, I'd be spoiled rotten — how'd three (tablespoons?) of nutmeg go in warm milk before bed tonight, with the scent of impending rain?" (Hold that thought, Mike, but also hold some of the nutmeg. Three tablespoons is a lot.)

Not all rain smells are pleasant. The ozone smell produced by lightning during a storm is not a favorite with everyone. In cities, acid rain interacting with organic waste and ground chemicals can give off some harsh scents. And acid rain itself, produced through the agency of automobile emissions and fossil fuels, can damage plants and water resources. When I lived in Louisville, Kentucky, our coal-fired power plants were killing fish as far away as Vermont and Canada.

What about the ability to forecast rain (or even snow) by smell? Lots of us think we can smell rain coming, or at least the freshening air ahead of it. Bear and Thomas, the discoverers of petrichor, give credence to this by suggesting that the smell released by oncoming rain on dry ground may drift some distance. Drought-stricken Australian cattle respond restlessly to this "smell of rain," they say. (My wife remembers that her family's chickens, not noted for intelligence, still went into the coop before a storm. "Chickens don't like to get wet," she tells me.)

There is one footnote to all this: if you type "Rain smells like God" into the internet and add the name Danae Blessing, you get 22,100 hits (as of 6/29/2006) on one of cyberspace's most popular stories, "The Smell of Rain" (original title, "Heaven Scent"). Briefly, Danae was born in 1991, months prematurely, and had a long fight for life, during which her parents, Diana and David Blessing, weren't allowed to touch her. When she was five, her mother remarked on the smell of approaching rain, and Danae replied, "No, it smells like him. It smells like God when you lay your head on his chest." (The story capitalizes the pronouns, but this is oral history.)

Various debunking websites have checked the story out and accept the authenticity of Danae's words, but one notes that "a leap has been made" between a five-year-old's remark and adult interpretation of this as a near-birth experience of God. Anyway, it adds, a preemie's olfactory cells couldn't have developed enough for smell to take place, an observation that might seem to underestimate the powers of the Lord of the Universe, if your thoughts are inclined that way. The debunking clincher: there are many stories of children who can see God or angels, but who soon lose this ability. Which, of course, Wordsworth knew when he wrote, "Heaven lies about us in our infancy."

* * *

As I wrote this, rain was moving in, and I took my coffee to the front porch to see what I could smell. A brisk breeze was blowing. The air was fresh and thunder was starting up. But sniff as deeply as I might, there was no hint of petrichor. Maybe it hadn't been dry enough. Maybe you have to be a thirsty cow in Australia.

6

WETTEST, DRIEST

HEAVY RAIN HAS FOUND ITS CHRONICLER in Binoo K. John, a freelance journalist and photographer in New Delhi, who traveled to the wettest place on Earth and wrote a book about it, titled *Under a Cloud*. The place is a village called Cherrapunji in the mountainous northeast corner of India, where more than five feet of rain once fell in a day.

Reviewers seemed bemused, when the book appeared in 2004, by John's fascination with such numbers. "I am supremely indifferent to the precise amount of rain that falls in Cherrapunji," Zafar Sobhan wrote in the *Daily Star* of Dacca, Bangladesh. And in the *Tribune of India*, Rajnush Watlas wrote, "The book's central experience is of singing in the rain." But both gave John generally good marks as a traveloguist bringing back reports of scenery and people in an obscure but fabled corner of the globe. "I suppose that is all one can ask for in a travel book," Sobhan wrote, sniffily.

It took a bit of digging, and some waiting, to get a copy of John's book through Indiaclub.com, but it was worth it. And in explaining his mission in the preface, John tells something about our fascination with weather records and our uncertainties about them. He writes of his mission: "I would look at life in the rainiest place and also try to unravel some of the mysteries attached to the phenomenon at Cherrapunji and the nearby village of Mawsynram where it rains slightly more."

So the rainiest place is not, in fact, the rainiest. Still other claimants, depending on one's source, are Mt. Waialeale in Hawaii and Lloro, Colombia. U.S. government meteorologists give their official nod to Mawsynram, with average annual rainfall of 467.4 inches. But John, who actually bumped his way up to Mawsynram by taxi over a flooded road, casts some doubt on the observations there. He reports these are taken rather primitively by a non-meteorologist named Pintoo. Until more sophisticated measurements take place, John writes, "it is Pintoo against the world."

While the U.S. government ignores Cherrapunji in favor of its possibly wetter neighbor, it is "Cherra" that history knows as "the wettest place on Earth." It was an important town to English colonials, who had a high rate of suicide there, and to the Welsh evangelists who followed them. It has had official meteorologists perhaps more reliable than Pintoo. One observer, B.L. Mondal, toted up what may be the heaviest rainfall in recorded history, on June 15, 1995, when Cherra got 61½ inches in 24 hours. During Hurricane Katrina, the Gulf Coast received 10 to 12 inches.

It is worth asking why it rains so much, and also why Mondal and his neighbors didn't wash out to sea. The answer to the first is that Cherra lies at the very end of the Indian monsoon, in a sort of meteorological box canyon where the clouds have to dump all their remaining rain. But because it is really a high plateau, not a canyon, the torrent runs off immediately and drowns Bangladesh on the plain below.

John writes poetically about life under the monsoon and avoids bathos, though sometimes just barely. He has a knack for pulling himself up in time with a wry comment, often at his own expense. One senses that he is not all that interested himself "in the precise amount of rain that falls," although he interviews weather watchers and writes up the statistics entertainingly. But he is diverted by everyone he meets, from the little girl selling cigarettes under a dripping sheet of polyethylene to the man sedated to keep him from leaving a hospital where he is the only patient. This is a little crazy-making for literary reviewers. One, who confesses to teaching postcolonial literature, writes, "I can't help thinking as I read it that the author had misconceived his entire enterprise."

I love the academic discomfort in that, because I think John did exactly what he wanted, which was to write a light-hearted, hopelessly journalistic, and at times lyrical account of a place where, for five months of the year, life is like standing under the bathroom shower with all your clothes on.

I am sorry to add that this carefree attitude seems to have gotten John into a wrangle with India's professional photographers, one of whom has blogged him nearly to death for selling his photos at a dollar apiece over the internet.

* * *

If there is competition for the wettest place on Earth, there is none for the driest. It is Chile's Atacama Desert, where NASA is doing research in a landscape resembling that of Mars.

Some parts of the Atacama have not seen a drop of rain in recorded time. Among places where it does rain, at least a little, the city of Arica holds the world record for lowest annual average precipitation—.03 of an inch. It would take 100 years to fill a teacup, someone notes. Irrigation allows the residents to survive, and indeed this northernmost city of Chile is a big tourist spot and calls itself "The City of Eternal Spring." Arica has better PR than its runner-up, Wadi Halfa in Sudan, which gets .1 of an inch of rain a year and is rarely visited by tourists.

Deserts are not as lifeless as they may appear. *National Geographic* points out that a million people live in the Atacama, in coastal cities and fishing villages, mining camps, and oasis towns. Astronomers love it because of the cloudless skies. But even in the desert's desolate core, life or its relics may not be totally absent, which is why NASA has been spending three years on a project called "Limits of Life in the Atacama," or LITA. One of the project's main goals has been to develop an astrobiology rover named Zoë, from the Greek word for life. The search for traces of life in an extreme environment—a Mars analogue—requires "long reconnaissance," moving fast and far to find scattered oases of microorganisms or their traces. Zoë does this superbly, even though it looks like something children might have built from tricycle wheels, a pine box, and an old push-lawnmower handle. It has what is endearingly called a plow, to skim off the top centimeter or two of soil. Sodbusters on Mars.

Zoë uses a "fluorescence imager" to spot markers of current or past organic life, and at first the instrument was overwhelmed by sunlight. The team has now developed a "gated" camera to screen that out. It will be used, the scientists say offhandedly, for "the upcoming search and specific study of habitats and life on Mars past and possibly present."

All their results were not yet in when this was written, although the second-year report noted that in the Atacama's arid core more organic molecules — DNA and proteins — were found than expected.

* * *

Dessication, desertification — the very words come drily off the tongue. The world's deserts are on the move, as they always have been. Ancient cities lie under Babylonian sands. But climate change and human activity may be speeding the process.

The U.N. has a Convention to Combat Desertification. Its website has a press release with a slightly startling title: "New ways to mainstream desertification." But this turns out simply to mean a move from planning to action, not a call for more deserts more quickly. And action is needed, to halt slash-and-burn farming, the squander of forests, and destruction of the fragile buffer zones between human communities and the desert.

One of those zones, the West African Sahel (a word meaning edge), was in the news repeatedly during the 20th Century's worst drought, from the late 1960s to the early 1980s. A quarter of a million people may have died, along with countless animals, devastating the economies of Mauritania, Mali, Burkina Faso, Niger, and Chad. On a website, I found an as-

tonishing photo of giant sand dunes bearing down on Nouak-
chott, the capital of Mauritania.

And yet that gives a somewhat misleading impression.
The desert does not arrive one fine morning, like a tsunami
out of a quiet sea. It encroaches a few feet at a time as farms,
forests, and brushlands degrade. Or it begins far away from
the main desert with the decay of small tracts. Drought accel-
erates the process but doesn't cause it. The Sahel crisis may
have begun with overfarming and the destruction of brushlands between farms and the Sahara, but it was spurred by the
failure of the West African monsoon. That failure in turn may
have been linked to slight changes in the temperature of the
Atlantic and Indian oceans.

The advance of deserts came under serious study in the
1930s, when drought and poor farming practices helped create
the American Dust Bowl and the Great Depression. Photos of
people from that time—by Walker Evans, Margaret Bourke-
White—do not look all that different from those of the Sahe-
lians. (Americans did not die in great numbers from famine,
though, and I have even heard a little Dust Bowl humor. A
man has fainted from the heat, and a friend says, "We had to
throw three buckets of dust at him to bring him round.")

* * *

All this is in my mind this morning as Hurricane Rita

moves across the Gulf of Mexico toward the Texas coast. Fueling its strength, the TV says, is a slightly warmer than usual Caribbean. Too much water in some places, not enough in others. "I think we can ride this one out," a Galveston resident says as he nails plywood over his windows. More than 6,000 people lost their lives in the Galveston hurricane of 1900, nearly 1,400 in Katrina during late August of 2005. Are we slow learners or what? How can we live on this planet in a way that doesn't destroy it or us?

* * *

I have never seen a real desert, but a biology-teacher friend visits one regularly to study plant life and take photographs. Once, he said, he had settled himself in a folding chair, gotten his camera ready, and then dropped off to sleep. When he woke, he said, he was surrounded by desert creatures — jackrabbits, iguanas — all watching him.

7

DRINKING THE RAIN

IT WAS AN IDEA to make glad the hearts of eco-freaks, and what's more it worked. In the early 1990s, the 300 residents of Chungungo in northern Chile began piping water down from giant mesh "fog collectors" on El Tofo Mountain. Suddenly the parched villagers had all the fresh water they wanted.

"Now I can wash every day," one of them told CNN. "Before, I had to watch every drop. You really suffer without water." Indeed.

I heard about Chungungo in a Sunday-school class where the lesson presented it as a shining instance of good stewardship of the Earth. The fog collectors looked like outsize sheets hung on poles to dry. Widely publicized, they led to similar projects or studies in 25 countries. But a few years later, Chungungo had almost abandoned them, and only nine of 94 collectors were still hanging. Most of the village's water supply was coming in, expensively, by truck. Early success had tripled the population to 900, and occasional water shortages fed the idea that the fog collectors were somehow unreliable. Village officials began agitating for a pipe to bring in steady water from a town miles away, even though laying the pipe would cost $1 million.

And there may have been something else going on here. Chris Smart, project officer for the International Development Research Centre, observes that "people have certain visions of what it means to be developed, and one of them is that water should be brought to you by the state and you should never have to think about it." Certainly you should not have to serve on the village fog-collection committee.

The story also points up the problem of distributing water equitably on our wet planet. Futurists have predicted that wars will someday be over water more than oil. And another kind of battle goes on within each of our bodies, which are three-fourths fluid and in constant search of the 2½ quarts each adult needs daily to survive in health. Dying of thirst—actually, of dehydration—is not pretty. In the soft-drink section of the grocery these days, I see my fellow shoppers in a new and more fragile way.

The amount of water on Earth and in the atmosphere stays virtually the same through the millennia—in no sense is Earth drying up. That won't start until a couple of billion years from now, when our sun becomes a red dwarf. Then it will happen in a hurry, geologically speaking. Meanwhile, the problems are ones of access, distribution, and keeping water clean enough to use. The Dead Sea is disappearing, not from drought, but from diversion of the Jordan River into farming and industry. Venice is sinking (though not so much now) because mainland industry drew off the pillow of water on which the city floats, like a child on an inner tube.

It can also be a problem that most of us take water for granted. In the 1960s, I edited a newspaper in Hornell, N.Y., where water flowed from my home tap reliably and at a modest cost. I eventually found there were no wells, pumps, or elaborate cleansing mechanisms. For a century, water had flowed by gravity from a reservoir in the hills to a now antiquated but perfectly good filter house, which passed it through sand tanks (with long German names) purchased abroad in the 1800s. When the operator needed a new part, he made one. Writing about the system, I crawled far back in a concrete tunnel, where—had I been a terrorist—I could have spun a wheel and left the town below waterless. My articles weren't well read, I suspect. Hornellians didn't care much where their water came from; they just wanted it there when they turned the tap.

The Chungungo process was even simpler than my town's, since mountain fog requires no filtering, and the nets that trap the fog are easily maintained. Some lessons were learned even in failure. The village water supply had not been the main focus, only the byproduct of a reforestation project. Bureaucracies were involved. The village had no history of civic cooperation. An organization called Fog Quest continues to work on fog collection projects elsewhere and is trying to resuscitate the Chungungo system.

In the human body, lack of water is signaled by thirst, our alarm for dehydration, which can quickly become serious and then horribly fatal. Hunger strikers who expect to be around long continue drinking, at least a little. Thirst, an encyclopedia explains, is controlled by osmoreceptors in the brain's hypo-thalamus, and "dehydration of the cells triggers the posterior pituitary to release the antidiuretic hormone ADH." Translation: urination declines; the body goes on a water-conservation kick.

One of the most lucid, and even funny, guides to avoiding dehydration is an article titled "Thirst and the Drinking Pilot" by Dan Johnson, a doctor who writes in *Soaring* magazine that thirst is a good danger indicator—"a reliable idiot light"—as long as you realize that it comes on only after you're already two to three percent dehydrated. "Is this clear?" he asks. "You lose your edge *before* you get thirsty. . . . [Thirst] is a sign that you should not get into an aircraft right now." You need to drink more, at once, to change yourself back "from a raisin to a grape." (Johnson also has some observations on what he forthrightly calls "pee collection" for pilots.)

As dehydration worsens, awful things happen to the body. At five to eight percent, Johnson writes, "you can probably still keep walking, but you'll have trouble figuring out what direction you're going." Survival stories from deserts and the sea describe the effects of even more serious water loss. The crew of the whaleship *Essex*, staved in by a whale west of Ecuador in 1820, suffered agonies of thirst and practiced cannibalism before the survivors were rescued. (Drinking untreated seawater and urine are not the answers, all survival manuals say, since these simply boost the body's salt content and speed dehydration. But you can survive for a time by sucking the salt-free fluid from around the eyeballs of fish.)

One of the most frightening dry-land accounts, "Desert Thirst as a Disease," was published in 1906 by a scientist, W.J. McGee, who described the "cottonmouth" stage of dehydration in which sufferers babbled water fantasies, even though "their brilliant ideas and grandiloquent phrases were but the ebullition of incipient delirium." And it gets worse. In August, 1905, McGee rescued a prospector, Pablo Valencia, who had crawled on hands and knees out of the Sonora Desert in Arizona after 6½ days without water. He should have died several days earlier. McGee, who had seen him in health, wrote in the *Interstate Medical Journal*:

> *Pablo was stark naked; his formerly well-muscled legs were shrunken and scrawny. . . . His lips had disappeared as if amputated, leaving low edges of blackened tissue; his teeth and gums projected like those of a skinned animal, but the flesh was black and dry as a hank of jerky; his nose was withered and shrunken to half its length, and the nostril lining showing black; his eyes were set in a winkless stare, with surrounding skin so contracted as to expose the conjunctiva, itself as black as the gums. . . . His skin [had] generally turned a ghastly purplish yet ashen gray, with great livid blotches and streaks; his lower legs and feet . . . were torn and scratched by contact with thorns and sharp rocks, yet even the freshest cuts were so many scratches in dry leather, without trace of blood. . . . We found him deaf to all but loud sounds, and so blind as to distinguish nothing save light and dark.*

We drink the rain. We depend totally on its ceaseless fall, its collection into rivulets, rivers, ponds, aquifers, rain barrels and reservoirs, and the canvas spread by lost sailors to catch an ocean shower. The balance is so durable yet so delicate. At good hydration, health. At a per cent or two below optimum, thirst, and at eight percent confusion. A little further, cotton-mouth delirium. Beyond that, Pablo Valencia.

* * *

A car goes by, with the driver's elbow carelessly out the window; her hand holds a plastic cup. A colleague arrives for lunch still clutching the outsize Diet Coke she brought from work. We kneel by a stream in the woods and scoop up a handful of the life giver. In a flood or a tsunami people die by the thousands surrounded by water useless or actively dangerous to them. We rehydrate relentlessly as long as we can, at any cost. If you were Pablo Valencia, how much would you pay for a drink?

8

THE SKY IS FALLING

IT HAS RAINED, according to one report or another, frogs, fish, tomatoes, blood, gobbets of flesh, fire, crayfish, coal, manna, straw and hay, worms, eels, cats and dogs, beans, maggots, gopher turtles, frozen ducks, cows, and the sky itself (see Chicken Little).

Did I say cows? Yes. Reuters is said to have reported in the 1990s that a cow, falling out of a clear sky, sank a fishing boat off the Siberian coast.

What in the world is going on? It's as if lovely ordinary rain weren't enough for us, and we needed disasters from the sky. Or that we're incorrigibly gullible, believing anything.

There's plenty of evidence for the latter in the study of strange rain, but there is something else, too. Some of these events are clearly miracle, legend, or hoax. Others have easy or at least possible explanations. But some fall at the interesting intersection of superstition and science, challenging us about what we know or can know, and throwing light on how our minds work.

To deal first with that falling cow. The explanation is that crewmen of a Russian military plane stole the cow from a field near their airbase. When it ran amok in the plane, they pushed it out a cargo door at 30,000 feet. Moscow reluctantly confirmed the incident, the news story reportedly said.

This is an airborne legend. Would you admit pushing a cow out of an airplane? Would your government? If there were ever a time for a cover-up, this would be it. (Why the idea of falling cows should be so compelling, or even funny, is beyond me, but I think Gary Larsen has a lot to answer for. His cartoons of shrewd, conspiratorial cows touch some dark strain in the human psyche.)

Also in the category of legends are rains of fire, which seem to be mostly Biblical and confined to very early times (the Cities of the Plain) and the Last Days. Dante placed blasphemers and usurers under such a rain in the *Inferno*. Some think Moses was talking to a volcano at Sinai, and modern rains of fire are indeed volcanic or, sad to say, are caused by humans flying at cow-dropping altitude. Incendiary war goes back at least to the Assyrians, and the Byzantine Empire used Greek fire—a sort of medieval napalm—to burn and terrify opposing navies. (Flying over the Soufrière Hills volcano on Monserrat, I saw what looked like a rain shower, but was really a fall of hot ash over the island's abandoned capital.)

Then there are the frogs. The Bible has a plague (not a rain) of them in Exodus, and a slimy business it is. But like manna, this is miracle, not meteorology. No scientific explanation of manna comes close to matching the scriptural details. There is a Rain-of-Frogs weblog on the internet, but this is whimsy. Frogs and fish do seem to appear at times of rain, but the usual explanations satisfy me — that a storm has sucked up a pond and its contents, or that frogs emerging after the rain have caused observers to leap (hop?) to a wrong conclusion. A Canadian friend once wrote endearingly to tell me that on a hike she had seen scores of frogs marching from one pond to another. "It was a sort of pilgrimage for them," she said.

Some other odd rains can be explained. Darwin thought red rain in the Cape Verde Islands was caused by Saharan sandstorms, and he was almost certainly right. Rain laced with microscopic diatoms also can look red. A report of a house plastered by tomatoes during a storm sounds possible, just. Mark Chorvinski has cataloged some of these events on a "red rain" website, although it puzzles me that nearly all his learned sources are from the 19th Century. Is no one doing weird-rain research these days?

It is beyond fish, frogs, and red rain that real doubt sets in. Rains of blood have been reported from antiquity, without much evidence, but in 1608 science finally got its chance. In July of that year, it began "raining blood" in Aix-en-Provence, a city in southern France. Red spots appeared on walls and the ground (though not on roofs), and local clerics decided these were the tears of God, shed over the exceptional wickedness of their parish. But they were called on this by a savant of the place, Nicolas-Claude Fabri de Peiresc, who had observed that a butterfly leaving a chrysalis also left behind a red stain. (That he called it "butterfly *merde*," as a modern-day poet suggests, may be a bit of literary license.)

The dawn of science began to dissipate what a student of those times has called "the fabulous gullibility" of the Middle Ages. But scientists sometimes brought their own naiveté. One of those who recounted the Aix events was Philip Henry Gosse (1810-1888), an English naturalist who was also the father of the noted Victorian poet and critic Edmund Gosse. The elder Gosse would seem to be a reliable source, but he damaged himself with scientists by his Christian fundamentalism and his theory that much of the evidence for evolution was a sort of hoax by God. Gosse proposed that at the instant of creation, God had also made the entire geologic and fossil record. Adam had a navel, he thought, because without one he would not have been truly human. Gosse was ridiculed for proposing "a God who deceives," and my 1911 *Encyclopaedia Britannica* says haughtily that he lacked "the philosophic spirit." But a recent biography, *Glimpses of the Wonderful* by Ann Thwaite, sheds a kinder light on him and may help restore his reputation as an excellent scientific observer. It was only in the art of reasoning that he was a disaster, she says.

With butterfly stains, we leave the realm of explanations and enter the territory of "something but what?" And here we quickly encounter Charles Hoy Fort, "the hermit of Brooklyn," who spent his life (1874-1932) culling tens of thousands of inexplicable stories from newspapers and journals — "anomalous phenomena," they're called today. Fort vouched for none of them and discarded none. He seems to have greatly disliked the certainties (and perhaps the arrogance) of science and felt he was rescuing "data" that might someday undermine its pretensions. He said some very strange things, as Forteans still do today. In the words of one essayist, Fort "refused to abandon the territory between science and the absurd."

He was quite interested in rains of blood, and of flesh. Such a rain in California, on Aug. 1, 1869, was commonly blamed on buzzards disgorging their prey, but Fort suggested it was just as likely that some living thing had been shredded during "teleportative seizures." (Fort is credited with coining "teleportation" long before *Star Trek*.) He had proposed earlier that rains of blood might come from giant creatures battling in space, or that the solar system itself might be a vast living organism that was hemorrhaging. No wonder early science-fiction writers loved him. But he never said he believed any of this—he was just collecting the "damned data," that is, data damned and rejected by ordinary science. A dirty job but someone has to do it, he seemed to be saying.

And now we come to the things, including some strange rains, that never were or for which data is scanty even by Fortean standards. Here one might place the medieval report that blood not only rained, but rained in the form of small crosses, and an Irish explanation of red rain as caused by mid-air battles among "the Little People." I was about to add a story of a coffin falling out of the sky, until my wife said she saw airline officials on TV, apologizing. I would like to believe all these things; legends are fascinating and I empathize with a friend who believes in the Philadelphia Experiment—that in 1943 the U.S. Navy found a way to make warships invisible. (If you also believe in this, please don't write to me. I will stipulate in advance that you are completely correct, and that I am part of a vast conspiracy to hide the truth.)

I also can't help liking Charles Fort. I'm a bit of a contrarian, too, and enjoy the periodic discomfiture of politicians, parsons, professors—and scientists. While it's not very likely that the solar system is a great sentient creature weeping blood on us, think of the doggedly held beliefs and theories that would be upset if it were!

Perhaps fortunately, some things are simply beyond proof or disproof. Many years ago, I read that the Prince of Wales had once reviewed a British detachment in Paris, where he lost his temper with a soldier and grossly insulted him (or maybe it was the soldier who insulted the prince). The report came from an old newsman who was determined to track down this suppressed story and who finally pinned it to a specific day and newspaper. He went to the paper's archives in a Parisian vault, found the volume, looked up the date, *and that day's paper was missing!*

Oh, yes, I believe that. I also believe in cows falling from the sky.

* * *

Rain fell today while there was very little traffic, and just enough water collected to make our street a mirror, with houses, trees, and passersby reflected perfectly in its depths, a duplicate world. Then cars began coming past again, their tires blasting the water off the surface and destroying the lovely illusion.

9

DEATH BY UMBRELLA

APRIL 20, 1814, WAS A RAINY WEDNESDAY in Milan, and this cost Count Giuseppe Prina his life, although he probably would have lost it anyway. Prina, the grasping Napoleonic finance minister for Lombardy, was dragged and beaten to death by a mob armed with umbrellas. It may be history's only case of an "ombrelletat," although conspiracy theorists think an umbrella figured in John F. Kennedy's assassination. More on that later.

The mob that killed Prina had been incited by news of Napoleon Bonaparte's abdication in Paris, and the illusory hope of Milanese independence. Earlier on the 20th, it had stormed the Milanese senate, where a Count Frederico Confalonieri reportedly set the umbrella motif by driving one into a portrait of Napoleon by the court painter Andrea Appiani, an act he later denied. The mob then went on to the house of Prina, who had shown a "singular skill in devising fresh taxes to meet the enormous demands of Napoleon's government," according to my 1911 *Encyclopaedia Britannica*. He was found donning a disguise and was "exfenestrated," another source says — thrown out a window.

Prina was able to make it through the streets and past La Scala to the home of a wine merchant who gave him refuge. But when the rioters threatened to burn the house, he gave himself over to them, saying, "Vent your anger on me, for I am already sacrificed to your fury: but at least I will be your last victim." The mob took him at his word, beating him, dragging his body through the streets for hours, and finally stuffing his mouth with papers bearing tax stamps. The *EB* scarcely concealed its disgust with this display of unbridled Italian *emozione*. "These horrors," it wrote, "were enacted by day, in a thoroughfare crowded with 'respectable' citizens sheltered from the rain by umbrellas."

There is a dark stain upon umbrellas, despite their often bright colors. A crowd moving under black umbrellas is an anonymous and unnerving sight. Jokes about umbrellas have an edge to them, often grounded in the human tendency to regard any umbrella in a rainstorm as fair game. In his old age, Ralph Waldo Emerson briefly forgot the word "umbrella," but easily made himself understood. "I can't tell its name, but I can tell its history," he told a memoirist. "It is taken away by strangers." A bit of doggerel by Sir George Ferguson Bowen makes the same point:

The rain it raineth on the just
And also on the unjust fella;
But chiefly on the just because
The unjust steals the just's umbrella.

It happens to all of us. One rainy day in Kyoto, my wife and I parked our umbrella outside the Nijo Castle while we inspected the "nightingale floors," designed to squeak under the feet of an assassin. When we came out, a woman was in the act of carrying off our umbrella. My wife thinks it was an honest mistake, but I take a darker view.

In Kyoto, we also formulated Rule One for sight-seeing in the rain: "Walk until your shoes squish, then hail a cab." Kyoto has only a bit more rain annually than Tokyo (62 inches to 60), and it is less depressing. A poet friend, Mike O'Connor, has written of "light-rain, small-lane Kyoto," where a woman once offered him an umbrella simply as a gracious gesture to a stranger.

Umbrellas (and parasols) are often things of beauty, especially in the Orient where they may have originated thousands of years ago. "Material offered by Mr. Du Feibao" on the web says a silk parasol of Hangzhou is "thin as a cicada's wing" and weighs only 8.8 ounces. It can be twirled fetchingly by the smallest girl and can also be used to maintain balance on a high wire.

Men's umbrellas were introduced to London about 1750 by Jonas Hanway, who had traveled in Persia. They were sometimes called "hanways," and hackney drivers taunted early users with "Frenchman! Frenchman! Why don't you call a coach?" The first U.S. umbrella (1772) was made in India, and on June 29, 1805, explorer William Clark lost his in a storm on the Missouri River. The first U.S. umbrella factory opened in Baltimore in 1828. In 1852, Samuel Fox invented the steel-ribbed umbrella as a way of using excess farthingale (petticoat) and corset stays. An African-American inventor, William C. Carter, patented the umbrella stand in 1885. An old household guide recommends cleaning especially dirty umbrellas with a linen rag dipped in unsweetened gin. And so forth.

A deep-belled umbrella offers more protection to one person and holds off the wind better. Umbrella sharers need a shallower one. Johns Hopkins students have designed an "anti-inversion" model, but well-made umbrellas rarely turn inside out except in cartoons. The elderly fellow in the background of the *Elmer* comic strip had neither umbrella nor parasol, and was always saying, "Durn this weather."

DEATH BY UMBRELLA

There's something atavistic about the way we hang onto our favorite umbrellas, even after the fabric has worn and the ribs have splayed. I owned one umbrella for 13 months in Taipei. Once I left it in a taxi as the driver pulled off, but I chased him on foot to the next light and got it back. I still have it, a nondescript tartan with a plastic handle that would hardly have bruised Giuseppe Prina. Someone will have to pry it eventually from my cold, dead fingers.

But still, but still—there's something sinister about umbrellas. In 1815, not long after Prina's death, a resident of Charleston, S.C., was murdered with an umbrella. Not for nothing does the Penguin, Batman's nemesis, carry a lethal one that can also lift him away from pursuers. The special horror of an umbrella as a weapon lies in its ordinariness, like the steel-lined bowler worn by John Steed in "The Avengers." (Steed's various umbrellas concealed a camera, a gas thrower, a tape recorder, and a sword.) Mary Poppins, like clowns, excites unease in many.

Which brings us back to the Kennedy assassination and The Umbrella Man, or TUM. Conspiracy theorists have analyzed the Zapruder film and espied (the right word) an open umbrella near JFK's car *on a bright sunny day.* Their argument, in essence, is that TUM used a CIA-supplied umbrella to fire a tiny flechette, or dart, into Kennedy's neck, paralyzing him so three riflemen could pot away. A fourth was unable to get off a shot. Truly, Horatio, there are more things in heaven and earth . . . .

The dark side of umbrellas may explain why there is also a comic side; at some level, we are still crouched in the cave, laughing uneasily at noises in the night. Bennett Cerf, in his jokebook *Try and Stop Me*, tells of a little man who listened quietly in church to a sermon on the Ten Commandments, until the preacher reached the seventh, forbidding adultery. Then he leaped up and exclaimed, "*That's* where I left my umbrella!"

* * *

It grew colder all day and threatened rain out of gunmetal clouds, which occasionally parted to let through a harsh and stagy light. But the redbuds insisted on starting to bloom, so sexy and abandoned that blossoms spurt directly from the main branches and even the trunks. Spring is a crime of passion.

10

JUST A GARDEN IN THE RAIN

IT WAS THE NAME—rain garden—that caught me, before I knew anything about such things. It sounded faintly Japanese, an image of rain falling through bamboo and perhaps a pool with water lilies and bright-colored koi. There may have been a whisper of my sister-in-law Nancy's old garden, where a wooden spout dripped slowly into a miniature ceramic rain barrel.

But Nancy's garden, beautiful though it was, was a mostly dry one. The words "rain garden" seeped into my mind from a newspaper story about Dan Welker, of Chestnut Hill, Pennsylvania, who had turned his boggy back yard into such a garden. A photo showed Welker on a wooden walkway, surveying his 40-by-50 feet of shallow ponds, marshy depressions, and water-loving plants. "There's not a blade of grass in Welker's yard," the story said. And, yes, there were water lilies and koi.

But the story disclosed a more practical and less romantic aspect of rain gardens. For many years now, we humans have been paving over the Earth. In many places, rainwater no longer soaks into the ground, which filters it into the aquifer. It runs off instead — quickly, sometimes violently — into storm drains, which send the water with its load of pollutants into rivers, lakes, and oceans. There are flash floods now in places that didn't have them 50 years ago.

My wife and I are not exactly strangers to runoff. The first house we bought turned out to be built on top of a ravine the developer had filled in. Stormwater continued to follow its old course, but this time through our basement. So the developer ditched and tiled around the house and installed a sump pump. This was fine, except that when the pump failed (which was often) the new tiles channeled *all* the outside water into our basement. It probably didn't help matters that I religiously tarred the driveway, assuring more runoff. In the town where we live now, all the houses seem built on ravines and the water table is high. Wet basements are endemic, but at least I've left my driveway graveled.

I had noticed, but not thought much about, the small markers that appeared a few months ago on the street drains of our town: "No Dumping — Drains to Creek," they said, with a picture of a frog. But of course every rainstorm dumps itself straight into them, carrying yard fertilizers, chemicals, oil, and dog doo into Hurricane and Young's creeks, our local streams.

When I began really thinking about rain gardens and storm runoff, I consulted my wife, who at the time was a newspaper reporter covering our city council and board of works. As I suspected, she knew a lot about stormwater. The drain markers, she told me, are in response to a federal mandate to reduce runoff, provide more public education, and totally separate storm and sanitary sewers. (Of course the feds have provided no money for this.)

Education is important, she noted. "People change their oil and dump it down the storm sewer. They flush their radiators and it goes into the sewer." She also hears residents demanding better, faster drainage. "If there's a swale on their land that holds water for a day or two, they're yelling mosquitoes and West Nile virus," she said. Mosquitoes only begin to breed in standing water after three days or so; the residents would do better to think about rain gardens.

A rain garden is just a way of channeling rain off roofs, yards, and driveways into the cleansing ground, holding it back briefly so nature can do its work. It can be as small as a dip in your lawn. "A rainwater garden is a strategically placed puddle," says Chris Cavett, an engineer for Maplewood, Minnesota, one of the rain-garden capitals of the world.

The idea flies in the face of a half-century of lawn propaganda—the idea of a perfect greensward, fertilized, barbered, chemically treated, free of dandelions and crab grass. "It's the only part of my life I can keep weeded," someone said. I had fallen for some of this myself. Dandelions and violets bloom freely in my scrappy, unfertilized, and chem-free yard, but I have a power mower that slurps gas at $3 a gallon and fills the neighborhood with racket. It also costs me $100 a year for mower maintenance. Not a blade of grass in that Pennsylvania rain garden, huh?

Across town, a couple of young women, obvious tree huggers, have ripped up their front lawn and planted flowers. I don't know if what they have is a rain garden, but its appearance was a shock to me. Doesn't a front yard have to be turf? Isn't there a city ordinance? But I've gotten used to it, and now check out the seasonal blossoms as I drive by.

The idea of slowing the runoff of rain goes far beyond home gardens. There are seminars on "stormwater management," and developers are required to build retention ponds in their new subdivisions. The more creative cluster their houses to reduce pavement and provide parkland. There are even "urban prairies" that filter rainwater through several stages into the ground and unpolluted streams. In cities like St. Paul, Minnesota, residents have banded to reclaim the forgotten creeks that once flowed freely through pioneer town-sites. "Daylighting," they call it, when part of a buried stream can be brought back to the surface. In Indianapolis, near my own town, a "public artist" has gotten a grant to paint a blue line through downtown, tracing the long-gone Center Creek that now courses through pipes under the pavement. The blue line is symbolic only, but it's a start.

The concern is a worldwide one on our watery but thirsty planet. India is exploring rain gardens and rain barrel technology. In 1998, Chengdu, China, opened a Living Water Garden, a six-acre ecological park in its inner city. A driving force behind the three-year project was Betsy Damon, who founded Keepers of the Waters in 1990 (in St. Paul, natch). The park, shaped like a fish, filters polluted water, and also features sculptures, an underground garage, an amphitheater, a forest preserve, and a teahouse. It all cost about $2 million. The park has succeeded, Damon says, even though the 80 multinational corporations that were approached gave a total of $6,000.

As an eco-freak wannabee, I applaud such efforts. But it's the idea of rain gardens that glamours me and may eventually lead me to build one in my own yard. How would a rain garden work for me? As I think about it, I realize it would start with a rain barrel—not great-grandmother's mossy, crumbling one, but a sealed 55-gallon drum (or a snazzier plastic one) into which I could feed a downspout. A barrel, or several of them, would address a problem my wife and I have contended with for years—not runoff, but run-in. Our downspouts dump rain close to our old house and it runs into the cellar, which is small, unfinished, and almost useless. The gas furnace and water heater are on blocks down there and the damp keeps homemade wine cool, but the floor is a muddy mess. We've tried plastic extenders on the downspouts, but they never stay put and have to be moved anyway for lawn-mowing. So stormwater management could start for us with a rain barrel.

But what then? At first glance, our yard looks peculiarly unsuited to rain gardening. The front yard is small and slopes quickly to the sidewalk. The side yard is a possibility, if the garden didn't disturb a huge holly tree that we're in love with. Where I would *really* like to take the water is to my flower and vegetable garden, 60 feet behind the house, across a driveway and a turf lawn. It takes two reels of hose to water it during high summer, and I spend a lot of time unrolling and rolling. Maybe I could pump water out of the rain barrel. Maybe I could hook a hose permanently to the rain barrel and let gravity do the pumping.

Now that I've become rain-garden conscious, though, I'm starting to think about that turf lawn between the driveway and garden. Most of it could stay—I'm not a lawnphobe—but there's a natural depression next to an ancient and grassed-over gravel drive. With a little digging and diking, I could have a small rain garden, watered from the sky or my rain barrel.

And in the rain garden could be some of those plants with the beautiful flowers and names—great blue lobelia, sweet-flag, queen-of-the-prairie, boneset, and that variety of stonecrop bearing my favorite of all wildflower names, "Welcome-home-husband-though-ever-so-drunk."

It's a dream, but at least I'm thinking about it.

* * *

I've come to depend on the rainwater pool that collects at the foot of my neighbor's driveway. A glance from my front porch tells me how heavy the shower is. It's also my laboratory for studying rain bubbles. When the weather clears, some of the water runs away down the gutter, but the rest stays for a day, gradually soaking into the disintegrating driveway pavement. If the driveway weren't in use, it could be a rain garden. But I know (I know!) that someday an asphalt truck will pull up, and my little pool will be paved so that stormwater (and pollutants) can rush away quickly into the drains and Young's Creek. It's the American Way—but does it have to be?

Photo by courtesy of Damaris Peck Reynolds

Flooding along the Grand Canal near Gaoyou. China, is shown in this 1931 aerial photo by Charles A. Lindbergh, who, with his wife, Anne Morrow Lindbergh, was on a mission to aid flood victims.

11

THE GREAT FLOODS

AS A CHILD, I NEVER FELT MENACED by floods. A memory from our summer camp in Indiana was of the high-water mark from 1937 on the front porch of our two-story house. I missed seeing that flood, but the idea of Sugar Creek coming up so high, with rowboats moored by our front door, was cheering. When we were on and in the water continually, why shouldn't the river itself nuzzle up to us now and then like a friendly sea monster? It was always called "high water," not a flood.

Later, as I grew up in Vincennes, Indiana, a river town, some of this insouciance continued. When the flood of 1950 came down the Wabash, I rode my bike through foot-deep water in intersections and got out of school to pile sandbags. It was hard work, but also a lark, punctuated by visits to the Salvation Army canteen truck for doughnuts. When sandbags had done all they could, the Army Corps of Engineers built mudboxes atop the city floodwall, and then someone saved us all by dynamiting a levee and flooding Illinois.

Other floods were lost in legend. There were 50-year and 100-year ones. Among the latter was the storied Midwest inundation of 1913, when residents of the prairie south of Vincennes were rescued in an Indian pirogue, carved from a whole tree and still hanging half a century later in a local historical museum. Farther back still was the "Great Freshet of 1847," which washed out roads and bridges across the Ohio Valley. The floods were mythic, but recently I looked up the death tolls; 732 were killed in 1913 and 250 in 1937.

Only with time did I realize that great numbers of people died in floods or had their homes and lives forever disrupted. And that these disasters were the products both of simple rain and of catastrophic events — earthquakes, bursting dams, typhoons and hurricanes, tsunamis. Not to mention God's wrath and the Deluge, the latter incomprehensible to a child, except that even then it seemed to me a dividing line between our world and something called "antiquity" — the word "antediluvian" resonated for me.

All knowledge ramifies. I now know there are deluge stories in many cultures, from the Babylonian to the Polynesian to the Native American. It's tempting to believe these reflect a real happening, although scholars say they probably are creation myths. Polynesians once thought the sky was a literal sea on which the sun and moon sailed. In straightening that out, the myth of a deluged Earth developed. A little nearer history, the stories linking Atlantis and a Mediterranean cataclysm before recorded history do not seem implausible.

The first historic flood listed in the World Almanac killed 100,000 Hollanders in 1228. In 1634, on the night between Oct. 11 and 12—St. Burchard's Day—a sea flood broke the North Sea island of Strand into fragments, drowning thousands. Sitting in a café on Nordstrand, one of the fragments, a German friend and I talked about the whelmed villages and legends of church bells still tolling under the sea. A few years later, my friend was dead, in a water accident.

People are alive today who passed through the greatest natural disaster in recorded history, the floods on China's Yangtze and Yellow rivers in August, 1931. Almanacks say 4.7 million people may have drowned, starved, or died of disease. Some estimates are lower, but does it matter? What news reports came out of that chaotic time sound curiously familiar. A million people evacuated from Hankow. Thirty million homeless across the country. Families gathered on rooftops in 100-degree heat, waiting for death and relieved only by waterworks employees who poled boats through the streets with tubs of water for them. An Associated Press reporter in a sampan approached an aged couple in water to their armpits, who begged him to kill them. Refugees reached Hankow in boats and floating on house doors and family coffins. There are some hauntingly beautiful aerial photographs of the flooding around Nanking, taken by Charles A. Lindbergh, who was in China at the time with his wife, doing relief work.

But there have been a thousand floods in China, and 1931 receded quickly from world memory, as the 2004 Indian Ocean tsunami with its 275,000 dead is beginning to recede, and as Hurricanes Katrina and Rita also will, more quickly than we would have thought possible.

What can we draw from this? That people build and live foolishly, too close to danger? This happens, but most of the 1931 victims were peasants whom the river sustained before it killed them. Where should they have been living? The St. Burchard's Day flood of 1634 swept away a system of North Sea dikes devised over hundreds of years, but the survivors built better ones. People are good at "risk technology." If we don't die, we live and learn.

In the United States, the Mississippi Valley flood of 1927 brought about the building of one of the world's great levee systems, which in turn hastened the disappearance of the wide marshlands that for millennia had soaked up the Mississippi floodwaters. There are obvious ecological drawbacks, but we tend to take our problems one at a time.

A commonplace after Katrina was that New Orleans was built in the wrong place and should be abandoned or moved. But writer George Friedman has pointed out that New Orleans is where it is because there has to be a port at just that place, or much of our commerce with the world fails. New Orleans will be rebuilt and repeopled whatever the cost. It has nothing to do with Mardi Gras.

Another commonplace is that we fail to respect nature, and try to conquer instead of co-exist. This is true, too, but we don't seem wired to take natural disasters lying down if we can do anything about them. In 2009, after 17 years of work, China is scheduled to complete the Three Gorges Dam on the Yangtze, the country's biggest construction project since the Great Wall and the largest hydroelectric dam in the world by a factor of two. The dam may end or ameliorate the Yangtze's terrible floods, while generating a ninth of China's electric power and opening the center of the country to ocean-going ships. The 400-mile-long reservoir may also silt up quickly or the dam split open—it is on a seismic fault, and an ominous crack appeared in 2000. Two hundred thousand are said to have died in an unacknowledged 1975 flood that began when a similarly situated dam collapsed.

The Three Gorges Dam may also end up costing $75 billion or more, displacing nearly two million people, and trapping toxic wastes in the river instead of flushing them out to sea. It will blot out priceless archaeological sites and matchless scenery. This might not be so necessary if China hadn't stripped its hills of forests over the centuries, or had maintained a system of inland lakes and wetlands. So many ifs, for all of us. If we were better people, if we really thought about our grandchildren and their grandchildren, if we were wiser, less greedy, not so rambunctious in the world. If.

The rain falls on us anyway.

* * *

Water is powerful in all its forms. In a moist woods in Brown County, Indiana, spring freshets bring down sizable trees. Under the rain, they begin decaying quickly. Fungus sprouts in a year or two, and after five or six years, you can kick easily through the trunks. In 10 years, the trees are gone.

12

DANCING WITH RAIN

WE BEGAN DANCING in terror and in guile—terror at the inexplicable world and a craftiness to outwit it. In the oak grove, by the sacred stream, we found those movements that seemed effective, and stylized them into gestures of prayer and propitiation. There is no evidence that woolly mammoths or saber-toothed tigers felt fear before the universe. It took our reflexive brains to realize we were in deep trouble and to begin looking for a way out.

How to live in the world without being at its mercy? Or the caveman's dilemma: how to be in and out of the cave at the same time? Humankind began dancing the answers very early. Its oldest gods were ones of weather—rain, storms, and lightning. Zeus was first a rain god. The Dogons of Mali, one source says, believe "that god's son the jackal danced and traced out the world and its future; the first attested dance was one of divination that told secrets in dust."

Serious stuff, then and now. In August, 1912, after spending a dozen years among the Hopi people, the photographer Edward S. Curtis was allowed to take part in their famed snake dance for rain. This was not a Sunday afternoon affair for tourists; preparation and the dance itself took 16 days. Onlookers didn't know, Curtis wrote, that "a white man was one of the wild dancers" who passed rattlesnakes back and forth and even held them in their mouths. Had the dance failed, Curtis's presence might have been blamed. So it was to his relief that "billowing dark clouds formed over the mountains and the welcome rain began to fall."

More recently, in 1998, fires were out of control in northern Brazil until, in desperation, the government flew in two Kaiapo shamans to perform a rain dance. Within hours, heavy rains extinguished 95 per cent of the fires.

Dance has been rooted since the beginning in weather, in rituals of rain and the marriage of Heaven and Earth. Gods spoke in the thunder and their seed fell on the Earth, fructifying it. Rainmaking magic is recorded on the earliest Chinese oracle bones; every Egyptian god was related in some way to rain. (Ramses II was a noted rainmaker.) The Mesopotamian symbol for transcendent height is "rainy sky." Dance has been, one writer says, "a sacrificial rite, a charm, a prayer, and a prophetic vision."

DANCING WITH RAIN

Sometimes the sacrifice was literal — children were slain to appease the strangely blue Aztec rain god Tlaloc. The more the victims wept, the more effective the sacrifice was felt to be. Sometimes it was symbolic — Arabs of North Africa threw holy men into springs to end a drought. But rain was always a matter of life and death, as it still is, even though most of us no longer believe it falls or is withheld at the whim of a deity.

The real origins of ritual are still there but hidden. Living in Taiwan for a year, I saw a lot of dragons without knowing their intimate connection to rain and fertility. The undulating crepe monsters that danced through the streets on Double Ten Day were direct descendants of ancient serpent deities, the spirits of ancestors now in heaven and sending beneficent rain upon their descendants. The Chinese expect results from their gods — a dragon that failed to produce rain might be whipped to shreds. "The Chinese are adept in the art of taking the Kingdom of Heaven by storm," wrote Sir James George Frazer in *The Golden Bough*.

And I find that I wrote something myself, out of ignorance, that was nearer the mark than I realized at the time. It was the story of a disconsolate dragon whose tears kept putting out his fire — until the tears were caught and sold to old bridegrooms marrying younger women!

Today our rain prayers, dances, and charms are likely to be similarly light-hearted. An encyclopedia of 5,000 spells has several for starting, stopping, and protecting against storms. To bring the rain, throw henbane into water, or uncooked rice into the air. Or place a terra-cotta bowl on your head for a few minutes, then fill it with water and bathe two cats. (This may stimulate more than rain.) For protection against storms, carry an acorn. To stop the rain, place a harrow perpendicularly at a crossroad. Or, if you are women in a group, throw water at a naked and premenstrual virgin as she dances. (Have a party," the instructions advise. "This is supposed to be fun.")

Rainmaking has gotten into popular fiction and even outer space. In Douglas Adams's *Hitchhiker* series, a rain god named Rob McKenna makes good money by promising *not* to visit your city. And out beyond Neptune is Plutonian Object 38628, an asteroid discovered in 2000 and named Huya in honor of a Venezuelan rain god.

* * *

Rainmaking in the last century or so has taken different forms, with different shamans. In 1850, James Pollard Espy proposed in a government report a scheme to produce rain over the eastern half of the country. He wrote:

Now, if masses of timber, to the amount of forty acres for every twenty miles, should be prepared and fired simultaneously, every seven days in the summer, on the west of the United States, in a line of six or seven hundred miles long from north to south, then it appears highly probable from the theory, though not certain until the experiments are made, that a rain of great length, north and south, will commence, on or near the line of fires; that the rain will travel towards the east side foremost; that it will not break up until it reaches far into the Atlantic Ocean; that it will rain over the whole country east of the place of beginning; that it will rain only a few hours at any one place, [and] that it will rain enough and not too much at any one place.

It's not clear why Espy thought all this rain was needed. The project was too big and might not have worked anyway; during an April freeze in 1910 orchardists in the Ozarks burned thousands of cords of wood and still lost their crop.

In the 1890s, "Melbourne, the Ohio Rain Wizard," did a brisk business at county fairs, sending gases into the clouds to provoke rain. And in 1912, San Diego hired a noted rainmaker, Charles M. Hatfield, to fill its reservoir. The ensuing rains washed out two dams, marooned a train, and led to the deaths of a dozen people. Hatfield was never paid, and the lawsuits went on until 1938. In 1930, the Belmont race track hired

George Ambrosius Immanuel Morrison Sykes, a "Zoroastrian minister," to keep its fall meeting dry. Sykes did, for the most part, although a final demonstration of his powers is said to have failed after pesky reporters filled his rain machine with clam chowder.

Scientific rainmaking began in the 1920s with the dropping of "charged sand" and dry ice into clouds and fog. Then in 1946, Dr. Bernard Vonnegut of the General Electric Laboratories hit on silver iodide crystals, which actually produced rain, if sporadically. The history of rainmaking since then has fluctuated, with environmental concerns tending to discourage large-scale climate-control efforts. There have been tragedies — a 1952 flash flood in Lynmouth, England, that some blame on cloud-seeding, and a 1972 flood at Rapid City, S.D., in which 200 died. The U.S. also tried producing heavy rain to flood the Ho Chi Minh Trail in Vietnam, with doubtful results. Just recently, the internet suggested that Al-Qaeda agents steered Hurricane Katrina toward New Orleans. We are back in the realm of myth and shamans.

And yet . . . . the ancient connection still holds and we still dance. In Taipei, I attended a performance of the Cloudgate ballet troupe in Taiwan's National Theater. At the end, dancers carried hundreds of lights onto the darkened stage until they merged with a backdrop of the galaxy. It was, as in the beginning, the marriage of Heaven and Earth, and I came out into the night awed and delighted and just a little scared.

* * *

Escaping the rain, I fled into the Deutsches Museum in Berlin with Carolyn, my co-leader of a student trip now washed out for the day. Carolyn walked on while I inspected a rough-hewn stone figure, eight feet high. "Hadad, Babylonian Weather God," a card said. "O great Hadad," I muttered, "do something about all this rain." And of course when we left the museum the sun was out. If you were a Babylonian weather god, trying to make a comeback, wouldn't you grant the wish of your only worshipper?

13

THE LANGUAGE OF RAIN

THE WORD "RAIN" APPEARS 101 times in the King James Bible, which is as good a place as any to start. The references range from the homely to the incomparable, from the great storms to "my speech shall distil as the dew, as the small rain upon the tender herb." There is the weather of Ecclesiastes, in the days of heedless youth, when "the sun, or the light, or the moon, or the stars, be not darkened, nor the clouds return after the rain."

Shakespeare also was no slouch at rain writing, giving us Lear on the heath and the Weird Sisters plotting their next meeting in "thunder, lightning, or in rain." Shakespeare's rain range is as great as that of anyone who ever tackled the subject, from Portia's "the gentle rain from Heaven" to "let the sky rain potatoes."

And there's Forrest Gump in Vietnam: "We been through every kind of rain there is. Little bitty stingin' rain, and big ol' fat rain. Rain that flew in sideways. And sometimes rain even seemed to come straight up from underneath."

We not only measure the rain, study it, drink it, listen to it, and walk out into it — we also imagine it, in an endless webbing of impressions, thoughts, words, scenes, and music. It would be absurd as well as impossible to list every great rain of the imagination. We all would have our candidates, not necessarily famous ones. In my favorite short story, "Passion" by Sean O'Faolain, the action turns on an Irish downpour that destroys an old man's beloved flowers. It is not an often-quoted passage, but is a perfect blending of plot and weather, with an intertwined love story.

Since rain touches the least of us, it's appropriate that a few matchless rain lines should come from a 16th Century writer whose name we don't even know:

> O Western wind, when wilt thou blow
> That the small rain down can rain?
> Christ, that my love were in my arms
> And I in my bed again!

How much of human woe is in those lines, and as in Deuteronomy it is the small rain that does it — not a kindly dew this time but the steady, slow mizzle of a dark night far from home.

The rain mizzles. It also falls, weeps, drizzles, drips, spatters, pours, streams, pitter-patters, drums, pelts, drenches, soaks, and sprinkles. It can be "a fine smirr of rain" (W.S. Graham), "the useful trouble of the rain" (Tennyson), or "the silver hosannas of rain" (Roy Campbell). It can also rain "like pouring piss out of a boot."

My favorite rain opening is "Drizzle becomes rain," which is how my friend Sam Dixon began *Outer Begonia*, his unpublished classic about a late coming of age in Taiwan. Hemingway (and I am pleased to mention Sam and him together) wrote a famous rain ending in *A Farewell to Arms*: "It was like saying goodbye to a statue. After a while I went out and left the hospital and walked back to the hotel in the rain."

Some writers and filmmakers use rain as a general atmospheric, without worrying much about nuance. You know the failed spy is going to jump off the bridge on a rainy night— how could it be otherwise? Gene Kelly, swinging on a lamp post, is a rare example in art of happy rain. More typical is Maugham's tale (titled simply "Rain") in which a missionary loses his virtue and then his life over a South Sea trollop, against the incessant drumming of the monsoon.

For the keenest observations of rain, one returns to the Bible or Shakespeare. How to explain the dread of "a little cloud out of the sea, like a man's hand"? Or how to praise enough

A FINE SMIRR OF RAIN

"They will out of their burrows like conies after rain"? Shakespeare was a country boy and knew his rain and his rabbits.

Not all good rain writing is in the classics, though. Here's a trim little poem written by McKenzie Burner, a tenth-grader at Plainfield High School in Indiana and published in the school newspaper, the *Quaker Shaker:*

> *The rain falls slowly,*
> *calms me in the long*
> > *dark night.*
> *What a smooth*
> > *talker.*

While writing this, I put out calls to several reading and film-going friends. We recalled Eliot's cruelest month "stirring dull roots with spring rain" and his summer "coming over the Starnbergersee in a shower of rain." We talked about the *Rains of Ranchipur* with Lana Turner and Fred MacMurray (!), and *The Rainmaker* and *Rain Man.* Rita Hayworth was remembered fondly. One respondent weighed in with *The Storm* by Kate Chopin, and "some film with a Wet Brando — was it *Streetcar*?" Another asked, "What was that French film in the 1950s where they're driving dynamite through a tropical jungle in the rain, and it explodes?" I have no idea, and it doesn't matter. There has to be an end to lists.

THE LANGUAGE OF RAIN

But there is no end to the imagination. In a poem called "The Changes," William Bronk wrote, "And there is weather here, and seasons," describing the world. Weather and change—light into dark, rain into sun, music into silence—seem built into the way we think and dream. My own imagination would fail in an endless succession of sunny days. I would be like the New England writer who was appalled as a child by a heaven where "there was no more sea."

And yet we don't know what it is, this ability to imagine and connect unlikely things; to listen to the small rain and turn it to a lament for home; to fashion words like "smirr" and "mizzle" and "ombrophobe"; to see in storm-shattered flowers the terror of passion (but it was not we who tied with a rainbow the moment when rain changes into sun). Why should we be wired this way, I wondered? I began to think about it, and not just in terms of weather. What peculiar things trees are, I thought—a fantasy writer might have imagined them, and here they are. This did not turn into an argument for intelligent design, but into something else—a sense of the persistent oddity of the world, that it should be this way instead of another. Finally, because I had to write something, I wrote:

Painterly clouds, and now the moon
is out there, right on time.
* * *

It's the way we would have imagined it;

A FINE SMIRR OF RAIN

hills the right size and the mountains
climbable, though tough. No monsters,
air about right so we aren't scared of it,
rain friendly for the most part.

* * *

Odd, really — just us, and clouds
at the right height, only one moon.

* * *

Rain, rain, go away — but come again another day. We have had
three days of fine fall weather, winey as the fallen pears from the tree
at the bottom of the garden. Already, I am eager for a change, want-
ing the cold rain and the darkening of the year, with fires of hack-
berry and oak, and the great winter festivals. And then the turn
toward spring again. I know a poet who used to take his little son
into the back yard at the winter solstice, where they banged pans to
bring back the sun. "It's all up to us," he said.

14

RAINY-DAY READING

WHEN IT COMES TO SCIENCE, I sympathize with the artist James McNeill Whistler, who failed the chemistry portion of his West Point entrance exam and said long afterward, "Had silicon been a gas, I would have been a major-general." It's not that I dislike science, disdain it, or have no interest in it. I simply lack the experimenting temper.

Which brings me to two books about rain—one a brilliant treatise on the idea of rain, and the other a delightful do-it-yourself manual by a born experimenter. I recommend each of them, for different moods.

A History of the Theories of Rain and Other Forms of Precipitation sounds ponderous and is not, although it does require some close attention by a serious reader. Its author is W.E. Knowles Middleton, D.Sc. and Fellow of the Royal Society, Canada. I knew I was in good hands when, after a brisk trot through the ancient Hebrews, Greeks, and Romans, I came to a short section titled, "The Next Fifteen Centuries." Middleton published in the 1960s, and his work may now be superseded in some respects. But *Theories of Rain* is a classic in the history of ideas, by a scholar whose good nature and wit peep out at the reader around the corners of the text. Take this passage:

> *René Descartes was a very great philosopher who had a tremendous effect on the progress of science, even though nearly all his detailed explanations of natural phenomena have turned out to be false.*

Or listen to Middleton as he turns aside after quoting Deuteronomy 32:2 and remarks, "How pleasant to say 'small rain' instead of drizzle." He also has a nice eye for chapter epigraphs, like the bit of conversation from Aristophanes in which Strepsiades asks, "Do you think Zeus always rains new water down, or does the sun draw the old up to be reused?" and Amynias replies, "I don't know and I don't care."

After his quick ramble through antiquity, Middleton gets down to business in the 17th Century with the invention of the barometer and the beginning of modern meteorology. (Middleton has also written a history of the barometer.) His 10 chapters, 200 pages in all, are neatly divided among ideas across the centuries about water vapour, the structure and suspension of clouds, electrical and chemical meteorology, and 19th Century theories of rain. He ends with 1914—this is history, not current news. There is a chapter on the barometer, and quick sketches on dew, hoarfrost, and hail.

Middleton's general approach reminds me of one of my English teachers, Rena J. Dunn, who promised to make me literate, but on the condition that I shut up and listen. Miss Dunn delivered, and so does Middleton. I took notes. For someone whose scientific knowledge is only superficially better than a 9th Century monk's, I now know a great deal about water vapour. I'm also incapable of Americanizing the author's spelling of vapour. There would be some terrible atmospheric disturbance if I did.

It would make no sense, and be impossible, to summarize what is itself a closely written summary. Read it yourself, perhaps on a rainy day. Middleton appears to have read everything ever written (although he says not) about what he calls "the hydrometeors." He carries his learning easily, and knows when to say, "We will now skip over . . . " or, "There is no need to pursue this matter further." The 17th and 18th centuries were wild times for meteorology, with outrageous guesswork, patient inquiry, brilliant intuition, and some flaming stupidity, to which Middleton is kinder than he might have been. Starting from the maxim that "vapour is not air," he leads the reader through a gallery of scientists, their theories, their experiments, and their mistakes. He never says too much and never patronizes. I had to look up a few words like "hypsometry" and "adiabatic," but they were worth it.

Even when Middleton encounters a fool, he is generous. He gives the personally difficult Jean André DeLuc his measured due as a scientist while noting DeLuc's "complete inability to criticize his own ideas." He does a masterful and good-humored demolition job on several other worthies. And of a pioneering paper, he writes, "a most elegant piece of scientific writing," which may stand as a description of his own prose.

My other recommendation is Duncan C. Blanchard, who wrote a little book, also in the 1960s, called, *From Raindrops to Volcanoes*. Blanchard worked at the Woods Hole Oceanographic Institution and is a raindrop freak—he would not mind that label, I think. His book, subtitled "Adventures with Sea Surface Meteorology," opens with an easy-reading chapter on "The Flights of the Raindrops." In fact his whole book is about water drops of various kinds, in the air, on the surface of the sea, at the intersection of boiling lava with ocean water. He gets in a fair amount of history, as well as much about the electrical side of meteorology. And experiments! If you want to tame a spider so it weaves a thread around a paper clip, Blanchard is your man. You will then be able to snag tiny drops of sea water on the thread and watch as they evaporate to salt. I skipped that, being a resolute non-experimenter and arachnophobe, but I'll bet his instructions work.

I first encountered Blanchard through his biography of Wilson A. Bentley, the Vermont "snowflake man." That book clearly was a labor of love and scrupulous in its details. Whenever I wandered from the prose, Blanchard's love for Bentley drew me back. And I suspect a poet in Blanchard. Here he is describing the microscopic fountain that erupts when a sea-water bubble breaks the surface film. The resulting film bubbles "bring to mind those magnificent rockets that are fired high into the evening sky to end many a Fourth of July celebration. At their zenith [this is one centimeter about the surface!] they explode into hundreds of pinpoints of light, to grow, and to expand into a twinkling canopy before they fall back into the water." Blanchard tells how to make them visible, and this is one experiment I would almost try.

There are many newer books than those of Middleton and Blanchard, more up to the minute, maybe even as readable. But if you get stranded somewhere in the rain, these are good ones to have with you.

* * *

Finishing these words, I went to check the mail and found the TV's promise of rain being redeemed on my front porch. The boards next to the steps are unprotected; the rain will rot them in the next couple of years. I could probably cover them in some way, but a little car-pentry every five years or so is no big thing. The wind is out of the north this afternoon, and cold, banging the wind chimes. But a few late cosmos are hanging on in the garden, and a brave geranium is still blooming in a protected corner of the side yard. I've never minded rain after the heat of summer. One summer there was hardly any for two months, and when a downpour finally came we ran out to the sidewalk and danced in it, feeling the hot cement go cool under our bare feet.

15

RAINY DAYS AND MONDAYS

I TOLD MY DAUGHTER-IN-LAW that I'd been walking around the house, humming old Carpenters' songs, and she replied, "You really are depressed, aren't you?"

The main melody I'd been humming was, of course, "Rainy Days and Mondays (Always Get Me Down)." There is a lively debate on the internet about whether "those silly Carpenters," as one contributor calls them, should have made it "Rainy Days and Sundays."

"Mondays are tough and all," he observes, "but nothing can make you melancholy like seeing your weekend drizzle to a close."

Michelle Saunders, writing in the Boston College student newspaper, asked readers to compare and contrast rainy days with Mondays. Mondays were popular with some on the ground of finiteness: "It can rain for a week straight, but there can never be two Mondays in a row," one respondent said.

This was contested by another who said, "Sometimes the rain will stop in the middle of the day, the sun will come out, and if you're really lucky there will even be a rainbow. But it is Monday all day long."

Most of those interviewed seemed to prefer rainy days to Mondays, and one even brought in Hollywood as evidence: "Good things always happen in the rain in movies. Long-lost loves are reconnected, a heart throb dies in a particularly gory death scene at the hands of an axe murderer [!], and women continually wear their white shirts outside and get caught in a sudden shower."

While there are a few happy rain songs ("Singin' in the Rain" comes to mind), more seem to be downbeat. One web site features examples of both in its "Ten Rainy Day Songs," which include, besides the Carpenters, The Cure ("Plain-song"), Garbage ("I'm Only Happy When It Rains"), REM ("I'll Take the Rain"), Leonard Cohen ("Famous Blue Raincoat"), Esthero ("Superheroes"), Blue Rodeo ("Rain Down on Me"), Cowboy Junkie ("Southern Rain"), The Doors ("Riders on the Storm"), and U2 ("Love Comes Tumbling").

Researching melancholy rain songs took me to *The Essential Waylon Jennings* as well as "Luckenbach, Texas," with its reference to "Hank Williams' pain songs, Newberry's train songs, and blue eyes cryin' in the rain." At this point I was out of my depth and called upon Jerry, my expert on old movies, country music, and auto racing. He responded with a downpour of e-mail suggestions, starting with Bob Dylan's "A Hard Rain's Gonna Fall," where rain is a metaphor for nuclear fallout, and Janis Joplin's "Me and Bobby McGee" ("Windshield wiper slappin' time"). Then he added "Who'll Stop the Rain?" "Raindrops Keep Fallin' on My Head," "Rainy-Day Woman," "Rainy-Day Women #12 and #35," and "Purple Rain" by Prince "or that symbol he used to use instead of a name." Jerry capped his researches with *The Shawshank Rebellion* and Tim Robbins in the rain with his arms thrust skyward. "This could go on forever," he concluded, "and leave me in a padded cell, and it would be all your fault."

It's clear that without rain, pop culture would need a new metaphor for depression. But what do scientists say about it? Most of their research has been on seasonal affective disorder, or SAD, a malady most often attributed to light deprivation. Clouds may contribute to this, but rain itself does not necessarily trigger SAD, which is a serious form of depression recognized by the American Psychiatric Association. It is more than the seasonal "winter blues" many of us get when the short, dark days arrive, aggravated in the mornings by daylight saving time.

"You see the SAD people beginning to go in the autumn," Dr. Norman Rosenthal, a pioneer in the field, has commented. "It's like watching the leaves fall off the trees. And then we started treating them with intense light exposure, and it's like seeing the tulips come up."

Another therapist puts it even more succinctly: "Fall comes and people get really goopy."

The accepted light-box therapy sends a flood of bright light against the retina and triggers messages to that part of the brain called the hypothalamus, which in turn revs up the production of serotonin and other mood lighteners. It also damps down the production of melatonin, a sleep-inducing chemical, by the pineal gland.

I asked Rosenthal if rain itself can induce melancholy, and he replied, "No doubt rain affects mood negatively in those who are susceptible, because it always comes along with cloud and tends to keep people indoors and away from the sunlight. I remember reading a wonderful quote by the Irish poet Synge, who talked about the rainy Irish countryside that was enough to drive a man mad."

Rosenthal added, "Interestingly, however, there may be other mood effects with rain. Some people may like the moist air, and the negative ions associated with rain may be conducive to a favorable mood in some (this is speculative, but there are data to support such speculation)."

A 1991 analysis by Rosenthal and others of SAD patients found that even after controlling for a seasonal effect, there are still day-to-day effects on mood, including cloud cover and barometric pressure.

In the popular mind, at least, there is little doubt of a rain-mood connection. John McManamy, who runs an internet site called "McMan's Depression and Bipolar Web," tells of moving to Vancouver to be with his fiancée and watching the autumn rains and mist close down the landscape. "They took away the mountains," he remembers crying and adds: "'I hate this city!' I screamed in the rain to my fiancée." Their romance was saved by moving to New Zealand.

"Theresa," a contributor to McManamy's site, says she keeps her Christmas tree up all year because she knows she won't have the energy to put it up in December. The site also describes a rarer "reverse SAD" disorder in which sufferers can't stand the warmth and light of summer. "Kathryn," writes: "I HATE spring, it fills me with foreboding." "Lynn," on the other hand, loves rain: "I wake up glowing," she says

Another site, on clinical depression, describes "rain gods" — patients who feel impelled to take responsibility for everything including the weather. It cites a woman who went into a tailspin of guilt when it rained during her garden party, despite the fact that sun had been forecast and that her guests moved into a large tent and had a great time.

Thought has also been given to weather and the workplace. A writer at the University of California Davis suggests replacing the old term "cabin fever" with "cubicle fever." He notes that on a recent rainy day his first thought was "I can't go out in this weather." At that point, he adds, "I realized that I had become a true Californian . . . . I have joined in the sense of entitlement to sunny or at least dry and moderate weather."

I consulted a friend, a mental-health worker, whose clients include SAD sufferers and who admits to at least a certain amount of sub-syndromal SAD herself. "It's almost like you carry around this rain cloud inside yourself," she said. But she also said her clients — many of them well-educated professionals — show up for their appointments better on rainy days. "Many of them have told me they knew they needed to be there on those days."

I also wrote to a Danish friend, Kim Nielsen, about rain and the winter blues in his part of the world. He replied that Denmark has a high suicide rate (about 9,000 attempts a year), but that the link to weather is unclear. In Greenland, he added, the suicide rate is about 20 times that of Denmark, which jibed with other accounts I had read. An Inuit resident commented that the problem was even worse in igloo days when "we all sat around together watching the walls melt."

Recommendations for avoiding the winter blues, and at least mitigating the effects of actual SAD, range from watching diet (to avoid carbohydrate binges) to exercising and getting outdoors in whatever sunshine there is. Positive thinking and following a plan seem to be helpful in themselves.

"If you truly believe," writes Andrew Solomon in *The Noonday Dragon*, "that you can relieve your depression by standing on your head and spitting nickels for an hour every afternoon, it is likely that this incommodious activity will do you tremendous good."

* * *

It's the third day after Christmas, and doing anything — especially writing about rain and seasonal affective disorder — seems like an insuperable task. The days supposedly are lengthening again, but who can tell? Yesterday was warm and rainy and one could imagine a hint of spring. Today is cold, dry, and overcast, with no hope. But then a late Christmas present, from our son's girl friend, thumps on the front porch. It's a pot with an amaryllis bulb just showing a little green — a successor to the lace-cap hydrangea she sent a year ago. With water and care, we'll have a flower in mid-February. And February is almost March, which is practically spring. Hooray!

16

THE WEBB FARM

THE WEBB FARM is going back to the rain. A 175-year cycle is ending, and a new one beginning, on 39 acres along the Shelby-Johnson county line in central Indiana.

Zachariah Webb was in his 20s when President Andrew Jackson signed a land grant on Oct. 5, 1835, giving him 62 and 58/100ths acres, part of it swampland. Zachariah and his wife Nancy built a house and barn, cleared and drained the land, and raised six children on it. They helped found a church, 2nd Mt. Pleasant, which exists today. Zachariah's father, John, had passed through Indiana on his way to Missouri; when John died, Zachariah rode horseback to Missouri to bring his mother back to the Indiana farm.

After Zachariah died in 1889, the farm went the way of many pioneer holdings, dispersed among another generation and outsiders. The Webb Farm is unusual only in that part of it stayed intact, identified with the original family, longer than others.

One of Zachariah's sons, Hampton ("Uncle Hamp") Webb, kept and farmed a sizable piece of it, bequeathing 35 acres in 1938 to his nephew, David O. Webb, with a life tenancy to David's widowed mother. After David's death, his wife Katie eventually passed it on to their daughter Sue. The deed describes the land precisely, in the way the continent was parceled out to its new owners: "35 acres off the north end of the west half of the northwest quarter of Section 2, Township 12 North, Range 5 East."

In the summer of 2005, Sue Webb—the great-great-granddaughter of the original settler—was on hand to see the Webb Farm, or a piece of it, dedicated as a protected wetland, developed by a new owner as a required offset for wetlands destroyed by the construction of a Wal-Mart a few miles away in the same watershed. What was a swamp in the 1830s is on its way back to swamp.

The 39 acres are a pastiche. Just over 29 were sold by Sue to the wetlands developer, who combined them with an adjoining 10 acres once part of the original farm. Sue kept five acres of "high ground." The site of Zachariah Webb's house and barn is on the land of Pam Parker, who sold the other 10 acres to the developer.

So the Webb Farm is going back to the rain. The ditches and tiles put in by Zachariah and his successors have failed; replacing them in today's economy no longer makes sense. Wetlands are needed, and the state has made it attractive for developers to construct them. Small "duck-diving ponds" have been built on the Webb Farm, and 3,000 trees planted.

Sue's mother Katie remembered a wood so thick on the 39 acres that "you couldn't see through it." Her husband Dave went squirrel hunting there with friends, and Katie carried breakfast to them. "I remember Daddy putting in more ditches," Sue Webb says; she also remembers a tornado in 1956 that felled some of the trees. She describes a slough on the land, which—being already a wetland—could not be counted in the offset for Wal-Mart.

On the afternoon of July 26, 2005, the 39 acres once belonging to Zachariah Webb were rededicated as a protected wetland. A marker was unveiled, and Sue Webb—a nursing instructor in Detroit—made a few remarks as a descendant of the pioneer farmer. Joe DeHart, the wetlands developer, also spoke, telling how he and his crew built ponds, set out trees, and planted native grasses. "We're hoping to draw back some of the native birds and the flowers they feasted on," he said. In eight or nine years, he added, the trees will have some size and deer will come here. In fact, Sue Webb said, she had seen one the night before.

The remarks were brief, and then, because it was Katie Webb's 96th birthday, everyone went back up the road to Pam Parker's house for ice cream and cake.

* * *

The dedication was on a fiercely sunny day with the temperature near 100. It's nearly the end of September and raining when I go back to the Webb Farm, phoning Pam Parker ahead of time so she'll know I'm not trespassing—although in a way I am. "We love to have people come and walk in it," she says. She doesn't love all-terrain vehicles, which have been a problem; in fact several teenagers were caught not long after my visit and hauled into court for driving their pickup trucks in the wetlands—Pam's husband spotted them, called the sheriff, and then blocked the exit until he arrived.

A few earlier tracks are visible as I turn off the county road and park on a small apron of crushed rock that Joe DeHart has left for visitors.

The rock runs out in a few feet, and soon I'm walking in deep clover along a raised track built to provide some minimal access to the wetland. A spatter of rain dents the surface of a slough to my right. There is little to see—just fields of tan grass and goldenrod, a few late butterflies, a grasshopper or two. Raindrops have globed on the clover, and soon my boots are soaked.

The track ends at a thicket of ash trees beside the remains of a wire fence that still clings to its rotting posts. The wetland seems very small — I can see to its diked edges in all directions, and the silos of a modern farm rise a quarter of a mile away. It is not even very quiet; dogs bark behind a screen of trees, a plane hums above the clouds. What would this field tell me if I knew it better, if I could read it? I step off the track and something curious happens. I'm now up to my knees in dried grass, which hides the uneven terrain. Suddenly I have to bring my vision in from the margins of the field to the compass of my next step. And that forces me to look at details.

A strand of tough spider web bars the path between two stalks of goldenrod. Pushing through is like breaking a telltale, and sends a yellow-and-black spider scuttling. Cattails crowd around, some of their smooth sausages now exploded into what look like hanks of dirty gray and brown wool. Deep in the grass are wild asters — white — and blue splashes of lobelia.

It's still raining and the footing is treacherous. What if I slipped and broke a 70-year-old ankle or knee? My wife would eventually drive out from town and look for me, or I could crawl back to the car, but what a mess — why didn't I bring a cell phone? Resolution: step carefully, don't break anything.

I move out of the deepest grass and take shelter under a shagbark hickory near the ash thicket. The hickory has dropped a windfall into the rotted heart of a fencepost, and the remnant of bleached nut looks out at me like a tiny skull. A Cornish author, A.L. Rowse, tells of being "pisky-laden" (pixilated?), lost in a field with no way out, in a landscape he thought he knew perfectly well. Something of that same feeling is beginning to impose itself on me. To be lost, 100 yards from a road. This field, this patch of wetland, is large when it can be traversed only by small, slow steps.

Taking those steps, I work my way around the back of the ash thicket and onto the entrance track again. I could walk quickly to the car now, but splashes of color in the grassland off to the right catch my eye — a patch of pink flowers and beyond them a spot of deep crimson. What are they?

Stepping carefully down off the track, I strike out through the grass to the first blossoms, which turn out to be cosmos, my favorite late-summer flower! I keep going through the grass to the crimson patch, also cosmos. And looking around, I see cosmos in isolated clumps like sentinels all over the field. Were they planted like this, or is it all coincidental?

Meanwhile, I've created a dilemma for myself. It's still sprinkling rain and I need to get to the car, the roof of which I can see 100 yards away. I could retrace my steps to the clover track, but decide to "triangulate" across the wetland. I begin pushing through waist-high grass, but the ground has been roughly plowed at some time and each step is a new consideration. After a few minutes, I look back and line up the crimson cosmos with a distant hill. I've been trying to plot a straight course, but instead am leaving a curving trail behind in the disturbed grass.

Now the terrain changes. The ground has leveled and become almost a mud flat, with sparse grass through which I can walk more easily. I stop for a moment and with my fingers scratch up a bit of the mud, a rich brown-black. No wonder Zachariah Webb wanted to drain this field for corn. Gloriosa daisies are everywhere. I look back one more time, but my cosmos has faded into the field. The car is only 20 yards away, and I push on through grass whose bowed heads make small nodding arcs. Just before I reach the car, a bird bursts out of the undergrowth, the first one I've seen today.

My walk through the wetland has taken a little over an hour, and I've covered at most 250 yards. It is still a landscape shaped by man, but in a few years it will be nearly itself again. Already I feel like an intruder. The weather is worsening. I climb in the car, back out onto the road, and leave the Webb Farm to the rain.

* * *

Tonight, the cold rain pools in my neighbor's driveway. It's the fringe of a hurricane, reaching us from the Gulf Coast. Not now, perhaps, but soon, rain will drip from an eave over the Ancient Rain Café in Taipei, and in the South China Sea a buoy will record a ringing bubble. Someone will photograph a cloud and date it. In Mawsynram, Pintoo will check his rain gauge. It will be a good night to sleep.

17

IN THE RAIN FOREST

MY UNCLE BILL first saw the great Congo rain forest on Sept. 1, 1946, while flying in a Lockheed trimotor from Coquilhatville to Stanleyville. He had finished his main job, securing three elephants for the Bronx Zoo in New York, and was off for a month of flying visits around what was then the Belgian Congo. Jotting notes in the plane, he wrote:

> *Clouds are opening — purple in every direction until it is lost in the distant clouds. This is the vast Congo rain forest they talk and write about . . . . My God! This purple never stops. We have been flying exactly one hour out of Coq & there is not a sign of it coming to an end. . . . Been running around and through a hard storm; the black clouds meet the black purple & the rain seems to whip up a froth on the horizon.*

I think of Bill's "My God!" reaction as I sit here in Indiana, on a cool October night, with a wood fire burning, and wrestle with statistics about rain forests. Some things are too big to grasp, and one is a forest that is 2,000 miles across. And is disappearing.

We all know it is, but our minds quickly become inured to this, as they do to a million AIDS victims in Africa. Even those most passionately dedicated to saving the rain forest fall back on absurd images: "Every second a piece of the forest the size of a football field disappears — 4,000 football fields every hour." With or without the end zones, I wonder. How many breadboxes is that? I took a picture once, in an Australian forest, of a strangler fig that had engulfed and eaten a whole tree. It was too big to get in the viewfinder, which is how I feel about rain forests.

The world's rain forests are a very big stand of trees, bigger than Winnie-the-Pooh's 100-Acre Wood or about anything else you can think of. Fifty million tribal people live there, as do 30 million species of plants, animals, birds, and bugs, which is another eye-glazing statistic, because who could ever count them? That 30 million is half the world's species. Earlier tonight I read, "In one Amazon tree stump there are more varieties of ant than in all the British Isles." I feel like Tertullian writing about miracles: "It's impossible, it must be true."

My uncle was a matter-of-fact man who liked to think he had a fairly ordinary imagination. So when he finally began hiking through the Congo rain forest, he didn't stay in awe very long. He wrote (in a newsletter prepared for his job as the zoo's curator of publications):

Once the jungle closed in beyond the radius of the village, its character seldom varied. Broad-leaved plants, endless vines, and trailing lianas, gray-green trees seen as shadows through the low bush but rising and spreading fantastically over-head — the jungle never varied, for mile after weary mile. In the full equatorial sun, crossing an airfield, the temperature was almost unbearably high, but not unbearably damp. In the forest, the shade was welcome, but the steaming humidity was like a hot, wet woolen blanket pressed against the body.

Still, it was exciting, interesting, stimulating at first — listening to the trill of birds, the hum and zzz-ing of insects, trying to keep an eye open for flowers or fruit to break the green monotony. But there was nothing — an occasional scarlet seed pod, a hacked-off pink blossom whose parent plant I could not identify, a few patches of broad turn-coat leaves, green on one side, scarlet-to-brown on the other.

It was always the same, green, green, more green varied by a leprous gray. The heat, the damp, the ants that fell on one's neck or swarmed across the trail and crawled, biting as they went, up one's legs — the fun soon went out of it, and the journey became a green nightmare.

A FINE SMIRR OF RAIN

Then I learned something of value; I learned that when you are on a tough trail like this one, it is best to turn off your imagination. Stop thinking. Don't look at the jungle any more, wonderingly, admiringly, fearfully. Just stop thinking and slog along, until the blessed moment when the bearers decide that they, too, have had enough, and slip off their packs for a ten-minute rest where the ants are comparatively few. During the breaks you can look at the jungle and find it exciting again. I remember, and always will, a garden of delicate toadstools, the exact color of moonlight, on one rotten log where I rested.

What that account does for me is break the spell of statistics. It is a description by a good observer of a real forest, not a romanticized one—you couldn't play football in it, but it's full of creatures like okapis and macaws and sloths, which move so slowly that tiny gray-green algae grow all over their bodies. It's also the home of the poison-arrow frog, a drop of whose venom smaller than a grain of salt can kill you. And of 2,000 plants with cancer-fighting properties. And of Toco Toucan, about whom my uncle wrote a book beginning: "Toco Toucan was sitting in a jumbie tree one day when Woolly Monkey came swinging along from one branch to another." He inscribed the book to me when I was five years old, writing on

118

the flyleaf: "Hey, Bill! I hope you're not too big to like a story about animals, because I wrote this one especially for you!"

If you absolutely must have statistics, I can recommend www.mongabay.com, a wonderfully comprehensive website run singlehandedly by a Californian named Rhett Butler, whose grandfather knew Clark Gable and who gives a damn (and then some) about rain forests. He will tell you that there are still 2.41 million square miles of tropical rain forest in the world — about five percent of Earth's non-frozen surface — although it is disappearing at an alarming rate. The overall Brazilian forest-cover loss in 1990 was 5,332 square miles. In 2003, after years of environmental agitation, it was 9,343, which fell to 7,298 in 2005. It is hard to believe that in 1911 the *Encyclopaedia Britannica* was reporting only 25 square miles under cultivation in the Amazon Basin. It takes a while to destroy a rain forest, but we're working on it.

Butler's breakdown of tropical rain forests by the world's four major regions shows the largest area, 1.08 million square miles, in Central and South America, the Neotropical region. Other major areas are the Ethiopian/Afrotropical, 720,000 square miles; Oriental or Indomalaysion, 390,000; and Australasian, 220,000.

I asked Butler how he keeps his spirits up, given the rate of destruction. He wrote back, "I'm not really a person who gets down easily. Deforestation in Brazil has fallen a bit this year, but I personally expect about 80 percent of the Amazon to be converted for agriculture or otherwise degraded. (Right now 18 percent has been cleared, while probably at least another 5-10 percent has been degraded.) While this is discouraging, there is hope that improved agricultural techniques — perhaps based on research into how pre-Colombian societies managed these forests — could maybe increase productivity on already affected areas and reduce the need for further forest clearing."

The Mongabay site explains why the destruction is happening — because peasants want land, because the First World wants Brazilian beef and mahogany, because legislators roll over for lobbyists, because Brazil needs to pay its international debt service. You can learn from the site about loggers, cattle ranchers, miners, oil drillers, land speculators, road builders, and slash-and-burn farmers. Maybe the farmers shouldn't be blamed too much. Rain forest soil is actually poor; all that verdure is supported by the quick change of rotting vegetation into nutriment, and farmers find their crops failing after two or three years.

You'll also learn about those who work persistently and patiently to try to halt the destruction — to keep alive these green lungs of the planet, so vast that they create their own rainstorms. But it's a tough job when a state governor gives out chainsaws as part of his election campaign, and when Brazil's annual budget for environmental law enforcement is only $9.5 million. It's hard to think about ecology when one mahogany tree brings $20,000 on the world market.

Butler writes, without needing any statistics at all:

The extinction event that is occurring as you read these words rivals the extinctions caused by natural disasters of global ice ages, planetary collisions, atmospheric poison, and variations in solar radiation. The difference is that this extinction was conceived by humans and subject to human decisions. We are the last, best hope for life as we prefer it on this planet.

Butler also wrote me that what Brazil decides to do with its rain forests depends a lot on "how Western countries value the services (especially climate moderation and biodiversity preservation) provided by forests. If Western countries start to place greater value on these services, then the protection of Brazil's rain forests can likely be 'purchased' via the open market.

"While right now the environment for such a scenario is not favorable, I believe it will become more so in the next few years. Scientists will play an important part in disseminating the value of these forests to policy makers and the media."

A FINE SMIRR OF RAIN

The rain forest troubled my uncle, a child of the white-bread American Midwest. In his Congo hut one night, he was awakened by terrible nightmares out of the primeval dark. At the time he was directing the Bronx Zoo's hunt for the rare and elusive Congo peacock, but things had gone wrong almost from the start, when the zoo's collector, Charles Cordier, fell and broke an ankle in the Ayena swamp. Bill got him out, trussed to a pole, but it took a march of two days, two hours, and 40 minutes to reach a road—Bill reported the time with great precision. He was glad to get home finally to New York City. But he would have understood as well as anyone I know why the rain forest matters.

* * *

Last night we drove home in the dark, in mist and fog, from the hills of Brown County, Indiana. Forests loomed around us, but it wasn't always this way. In the 19th Century, pioneer farmers stripped the hills and for a few years tried to make a living on the rocky and rapidly eroding slopes. Then they gave up and left. Today a friend who owns 40 acres of forest has what may be the state's biggest black-cherry tree. This kind of forest can come back. The question is whether the vast ecosystem that is a tropical rain forest will be able to.

18

THE TROUBLE WITH VENUS

THE TROUBLE WITH VENUS was that it could never rain there. Earth's neighbor had a runaway greenhouse-gas problem. As it got ever hotter, the atmosphere absorbed more and more water without reaching saturation. All that water eventually blew off into space. There is no water on Venus today — its lovely sparkle comes from deadly clouds of sulfuric acid.

Even the most devout believers in global warming don't see Venus in our future, at least not any time soon. But there is a widespread conviction among scientists and others that we're on a risky path if we fail to limit gases like carbon dioxide. Opposing them are organizations, mostly conservative and business- or consumer-oriented, who argue that the cure would be worse than the disease. In fact, a little global warming might be good for us, they argue, by making northern lands more hospitable. But it may be a moot point, because "there is no serious evidence that man-made global warming is taking place," the National Center for Public Policy Research said in 2004.

It all has a familiar ring. Thirty years ago, as an editor at the *Louisville Courier-Journal*, I coordinated a series of 20 newspaper articles on the history, economics, and dangers of tobacco. The Tobacco Institute was arguing vociferously that there was no serious evidence of a link between smoking and health. We found a link, but it took many years and lawsuits to establish the danger absolutely. Now I watch anti-smoking ads from Philip Morris, which has diversified. "There is no such thing as a safe cigarette," the ads say. (The vilest U.S. tobacco product in the 1970s was "black fat," made from the stems and other waste parts of the plant. It could be sold only outside the United States. Googling "black fat tobacco" recently, I turned up an offer from someone in Nigeria to buy "black fat [burley] or its equivalent. Color black with nice scent. Submit price quotation please.")

Global warming is not tobacco, but the tenor of the arguments is similar, and discouraging to anyone seeking a definitive answer. So when I went to my local library and saw a book titled *Is the Temperature Rising?* it sounded like a quick way to find out. It wasn't. Instead I got an engrossing and beautifully written textbook by E. George Philander, professor of geosciences at Princeton University. It was published in

1998, and a subtitle reflects the tentativeness of only a few years ago: *The Uncertain Science of Global Warming.*

Philander makes a startling statement early on: that no matter how much greenhouse gas we pump into the atmosphere—whether carbon dioxide, water vapor, or something else—it will probably not destroy the planet. The Earth has rebounded from catastrophes before. But it may not be nearly as pleasant an Earth for us. While some see little danger, he says, "some experts are warning us that we are making poor bets" by deferring action until a final scientific verdict is in.

What Philander states unequivocally is that in 1998 we were pumping six gigatons of carbon into the air each year and "increasing the concentration of several greenhouse gases, not by a small percentage, but by a factor of two or more." Doubling our bets, that is. Six gigatons is six billion tons. Half of that stays in the atmosphere, according to Philander, and we're not sure where all the rest goes—except that as we cut down the forests, a major "carbon sink" disappears. The growth of greenhouse gas is exponential. If you're in a boat headed for a waterfall, Philander suggests, it might be wise to pull into the bank before the current gets much swifter.

Philander appeals to me because he writes with a total lack of hysteria, and describes—sometimes lyrically—the cycles upon which our life on Earth depends, from the way rain forms to the motion of the winds to thousand-year currents in the deepest abyssal oceans. To step from Philander to the internet is to travel several years and a psychic distance as great as that between a Zen master's hut and Times Square. The debate over global warming is in full cry on the web, spurred by several very warm years and the 2005 hurricane season. Hurricanes feed on warm water, and tropical oceans are about a degree warmer than they were in 1970. Alaska had one of its warmest summers on record, with gardens flourishing into October. WorldWatch reports some dramatic melting of glaciers and polar ice. The U.S. produces a quarter of the world's carbon-dioxide pollution from fossil-fuel burning; a start on a solution, we're told, would be to limit auto and power-plant emissions sharply. Jim Hansen, director of the NASA Goddard Institute for Space Studies, writes in the July 13, 2006, *New York Review of Books* that we have about a decade to do something that will at least mitigate the crisis.

The other side argues that any warming is cyclical, a normal part of nature, and not evidence at all. "Global Warming-Hurricane Link Just Hot Air," says a press release from the National Center for Public Policy Research.

Still, it doesn't take a weatherman to see which way this particular wind is blowing. Googling "global warming" turns up 127 million hits, but those at the top, the ones with the most

"Google juice," are those sounding the alarm. The faceoff is largely between scientists and environmentalists on one hand, and those for whom economics trumps a possible future problem — between those who want to get the boat to the bank now and those who think we can ride a little further, if in fact a waterfall is even there. It does not help my faith in the latter group that a substantial funder of the National Center for Public Policy Research is the Exxon Foundation.

In his book, Philander barely answers the question in its title: Is the temperature rising? It is "very probable," he says, that the average global surface temperature will rise between .9 and 3.6 degrees Fahrenheit by 2050. Sea levels are likely to rise five to 50 centimeters, or two to 20 inches. These figures may not seem like much, he notes, but a swing of 10 degrees triggered the last global ice age.

Philander also tells a little mathematical story, about a gardener who finds the number of lily pads in his pond doubling every day, from one to two to four and so on. On the 100th day, the pond is full, but on what day was it half full? The answer is day 99, and the moral, of course, is don't wait until then to do something about it.

* * *

When I was a child in school, we had a primer titled "Fifty Winters Ago," which would have been, at that time, the 1890s. The scene, as I recall, was a New England paradise of snow and lighted farmhouses and horses that pulled jingling sleighs. I wondered even then why we never seemed to have such cold and cozy winters. Some of the predictions of global warming are based on no more solid data than my recollection. But then there's the World Glacier Monitoring Service, which is so relentlessly scientific that I can barely understand when it talks about glacier "balances." These seem to be something like bank balances for ice: how much has been gained or lost since the previous year. What jumps out is that of 60 glaciers, the balances for all but five are negative.

19

THE ICE MINERS

IT WAS RAINING ON A MORNING, unimaginable ages ago, over the spot a race of short-lived creatures would later designate as 75°06' S, 123°21' E.

But now the atmosphere's gradually falling temperature was beginning to change the rain to snow, dusting the bedrock at first and then clothing it permanently in white. The snow went on falling as the rain had, burying the land that the short-lived creatures would call East Antarctica. Immense pressure and cold would turn the snow to ice that after more ages would grow to be 3,275 meters thick—more than two miles—at that particular spot.

On another morning, the short-lived creatures came to 75°06' S, 123°21' E and set up a huge hut with a rounded roof, on a geological feature they called Dome C. Then they began drilling. They drilled for nearly 10 years, bringing up a worm of ice that eventually reached almost to the bedrock where the first flakes had fallen, some 900,000 years before. To the creatures, 10 years was a long time—one-seventh of their lives. And even though 900,000 years was itself a tiny fraction, perhaps 1/5000th, of the life of Earth, it seemed so immense that they used a special notation, 900 kyr, to make it easier for their minds to grasp.

The last ice they brought up was from five meters above the original bedrock. Then they stopped, because the Earth's heat was melting the ice at the rock face. This last core was older than their species itself.

* * *

It sounds like a science fantasy, but it's true. The story is of EPICA, the European Project for Ice Coring in Antarctica, which began in 1996 and ended in 2005. Scientists of the project—a consortium of 10 nations—examined the air trapped within a two-mile-long cylinder of primordial ice. It was the actual air of antiquity, not just some footprint of it. Those at another drilling site studied a shorter ice core in detail. Both groups built on the knowledge gained from the earlier Vostok Antarctic ice core, as well as extensive drilling elsewhere.

Scientists have been coring the Greenland ice cap for decades, and one aim of the Antarctic drilling was to study the climatic linkages between the hemispheres, and how these may have changed over time. This polar ice mining is part of a broader effort to portray archaic climate throughout the globe. An ice core from a Bolivian volcano has given a picture of tropical weather over the last 25,000 years. Another, from the Tibetan plateau, may go back 200,000 years or more. And three cores from a Himalayan glacier are the highest ever re-trieved—from a work camp at 23,000 feet. (The cores, glisten-ing tubes several inches in diameter, were cut in sections, manhandled down the mountain, and carried by yak to the nearest truck road.)

Ice mining is a tough, gritty, and cold job. The drillers don't get poetic about it, any more than coal miners do. Oper-ating their very specialized drills has its hazards. Ice miners in Greenland have fallen down crevasses and died. Drills have touched bedrock and frozen to it. (The Dome C drill got stuck in its shaft in 1998, delaying the project while a new shaft was sunk.) The one- to three-meter core sections were handled with gloves and shipped to a frozen-food warehouse near Grenoble, to repose, amid goat cheese and frozen raspberries, until they can be analyzed spectrographically. The ice is fairly clear, since its air bubbles have been drastically compressed— "have had the breath squashed out of them," a writer says.

But the bubbles are still there, and from the reclaimed air scientists have learned that atmospheric carbon dioxide (a greenhouse gas) and temperature have been marching in lock-step for hundreds of thousands of years—a finding of interest to students of global warming. From the ice itself, researchers can tell the temperature of frozen rain that fell nearly a million years ago.

The Dome C core has also doubled, from four to eight, the number of ice ages for which detailed records exist. The earlier four, it found, had longer and cooler interglacial periods than the more recent four—in short, the difference between warm and cold periods was less extreme than it has been since a change toward wider swings began about 450,000 years ago.

The findings supported those of marine scientists, who analyze climate through deep-sea sediment composed of the shells of the microorganism *Foraminifera*.

It was hoped the Dome C core would also contain evidence for the last reversal of Earth's north and south magnetic poles, 780,000 years ago. (We are overdue for another one, and no one is sure what, if anything, will happen.) But the "ice-core community" has been frustrated, says Dr. Jerry McManus of the Woods Hole Oceanographic Institution, because a beryllium isotope "spike" that would mark the reversal hasn't shown up so far. (There is no "magnetic recorder" in ice as there is in deep-sea sediments.)

THE ICE MINERS

Some results of the Antarctic studies were published in 2004-5 by the journals *Nature* and *Science,* under titles that do not invite casual reading: "Eight glacial cycles from an Antarctic ice core" and "Atmospheric methane and nitrous oxide of the late Pleistocene from Antarctic ice cores." But they repay careful attention. One cannot read them without awe at the ingenuity of the short-lived creatures. Even the barest acquaintance with glaciology introduces one to a radical vocabulary: "mid-Brunhes event," "Milankovich effects," "marine isotope stages," "Dansgaard-Oeschger events," "Termination V." The Dome C core was expected to be the longest continuous one that could ever be retrieved, reaching the "mid-Pliestocene revolution" when our current series of ice ages and intervals began. But in late 2006, scientists said they might be able to find a core going back a million and a half years.

It's difficult to comprehend such things and most of us don't need to, although they have their fascination, and a day of internet research makes them understandable, at least in outline. The terms are a specialist's language for talking about ancient and future environments—how these are influenced over millennia by the Earth's obliquity and orbit around its star, by its slowly gyrating poles, and by volcanoes and the grinding of continental plates. "We live in an old chaos of the sun," the poet Wallace Stevens wrote.

A FINE SMIRR OF RAIN

One comes back to the ingenuity of the ice miners and their cohorts — the wit to chart, at first from ancient dust and tiny corpses in the sea and then by drilling two miles into the ice, the weather of a morning nearly a million years ago. The ice miners tell us things. That ancient annals of extreme cold are true, sometimes to the precise year. That the historical analogue for Atlantis may indeed have perished in a volcanic cataclysm. That we are living in a mostly benign time, climatically, which has lasted 12,000 years and may go on another 12,000 or more before the ice returns, if we don't disturb the balance too much. (But a writer in *Nature* observes that "the predicted increase in greenhouse-gas concentrations makes this unlikely.")

* * *

Dansgaard-Oeschger events are rapid climate fluctuations that occur during and at the end of ice ages. Willi Dansgaard, a Danish scientist, shares discovery honors with the late Hans Oeschger, a Swiss glaciologist. Dansgaard is one of those pioneers described earlier who flew in small planes through thunderstorms, gathering raindrops. (His wife went with him, she said, "because I didn't want to be a young widow.")

Dansgaard has written an entertaining memoir, *Frozen Annals*, about ice mining. He tells how he realized, in the 1950s, that certain isotopes, or variations, of oxygen molecules could be used to chart temperature and other phenomena over time. He began researching the idea in his yard, using a funnel and a beer bottle to collect rainfall. Later he spent a career applying the concept of "stable isotope analysis" to ice cores from Greenland. In *Frozen Annals* he writes that it "was a good idea, maybe the only really good one I ever got." What he doesn't say is that this is one more than most of us.

* * *

From my front porch, on an early winter day, I can watch rain becoming ice. This ice is perishable and beautiful, especially when it clothes the trees in a glittering mantle, or forms a pane on my backyard pool that I can pick up and peer through like a window. Dangerous, too, for a short-lived creature who has 70 winters behind him (.07 kyr in the language of glaciologists) and has learned through many falls to negotiate the ice with care. But ice also preserves and protects, at least briefly. A few years ago, I went to see the Ice Man, dug out of a Tyrolean glacier in 1991. His mummy was on display at the National Geographic Society in Washington, D.C. We looked at each other across 5,300 years or 5.3 kyr — an eyeblink compared to the 900 kyr at the bottom of an Antarctic ice mine. And less than that to the 4,500,000 kyr age of the Earth. Or the 20,000,000 kyr that my long-outdated atlas tells me may mark the beginning of the universe and of time itself.

A FINE SMIRR OF RAIN

20

WRITTEN IN WATER

Who can number the sands of the sea, and the drops of rain,
and the days of eternity?

    —The Wisdom of Jesus, Son of Sirach (Ecclesiasticus 1:2)

AS WITH SO MUCH WRITING, this book began as a pastime and a probe and then turned serious. "I write to see what I'm going to say," Annie Dillard wrote, but for a long time the direction was uncertain. Was I looking at some vast metaphor? I hoped not—vast metaphors are not a strong suit of mine. Gradually I realized this was not metaphor—one thing standing for another—but synecdoche, in which a small piece represents the whole. The world was too big a subject, but I could write about a raindrop.

    The writing began in March, 2004, and went on intensely through that spring. I planned to keep a daily rain journal, but the impulse flagged. Instead I was writing down fragments that might fit somewhere, but didn't always; I never found any more about Foc Foc on the island of Reunion where it rained six feet on Jan. 7-8, 1966—more, apparently, than at Cherrapunji, but perhaps the measurements were flawed. I read about the mapping of Mars and its great outflow channels from some primordial deluge, as if "all the Great Lakes had emptied into the Mississippi in a couple of weeks," a source said.

As the book neared its end, I became a volunteer copyeditor for Rhett Butler on his rain-forest website. One night we calculated during an e-mail exchange how many humans might live during the five million years it could take to fully restore the rain forests. It came to 500 trillion, give or take a few trillion, an absurd figure.

I had some concerns about whether I could ever draw a complete picture; isn't all writing a fragment of some larger whole that we may not even identify? I wrote down Etienne-Jean Delécluse's comment on Stendhal's intelligence and its gaps: "It is like a bullet hole in a flag. The flag keeps on flying, but through part of it you can glimpse the sky beyond." I liked the idea of being able to see through the writing to some wholeness. And this also suggested a Buddhist sensibility: "We should always live in the dark empty sky."

In the summer of 2004, unusually wet worldwide, I wrote to several friends to ask about their memorable rains. Their answers, like differently angled mirrors, showed me the subject in new ways.

Linda Yew, a missionary in Kuala Lumpur, replied: "My most vivid storm? The Life Storm. By the Grace of God, I've gone through a few and come out a stronger person." The natural storm Linda remembered best was her first typhoon in Taipei, when wind howled at her door "like a woman wailing in deep sorrow."

Tammy Peng, a lawyer and mother in California, told me how she sat with her new baby, listening to rain. "It is critical to find out if the infant has any hearing disability. So I would just sit there with her, talking to her, teaching her to listen to the rain and to appreciate it. This went on for a while until she turned toward the windows where the raindrops came together to make music. Then, I knew, my baby could hear!"

In Scotland, my inquiry reached Ann Wakeling as she was preparing to ride her bike the length of Britain, from Land's End to John o' Groats. Ann kept a few rain notes for me as she cycled: "The rain pings on my helmet, and then it goes through all the holes and my brain goes soggy. Then I realize that the new 'waterproof' gloves (made in a hurry) have not had the seams sealed, and are leaking, and the inner glove is getting damp, damper, wringing out sort of wet and uncomfortable. . . . Misty rain on spectacles renders them opaque, difficult to see clearly where you are going." Ann later forwarded to me a poetic report from her son Neil on the particularly sodden Scottish summer of 2006: *Wet / wind and wet / really heavy wet / even heavier wet / really really heavy wet with wind / gales and wet / wet and wet.*

Mike O'Connor, a longtime rain watcher, wrote to remind me of the sound of rain on banana leaves in Taiwan — and that the leaves (like those pictured on Page 144) are designed to split without injuring the plant when rain beats on them. Mike also let me quote from a chapter on rain in his memoir *Montesano*, about growing up in the Pacific Northwest:

> *Aside from a population of crazy children, the rain, our teachers told us, grew the biggest trees in the world! And indeed the trees do love the rain. Their roots rejoice and their buds burst and flower. I don't know where all the birds went when it rained for weeks. I know where the ducks went; they took over the Elma Country Club golf course. But it all came down to this: I was lucky and grateful to live in a house with responsible, if imperfect, parents, a sister who was studious and mostly stayed out of my way, books and oil heat, and plenty of Mason jars of canned fruit in the basement. And there was no better way to endure the rains, the weeks of rain, the months of rain, and even in time to come to love what could be called the Republic of Rain, than to be at*

home — or to be able to carry "home" around in the wet and often stormy world with you when you weren't.

Finally, in some old letters, I found that my Uncle Bill had written vividly about rain in the Congo in 1946. Flying out of Aba, in the country's northeast corner, he wrote that his small "ratty" plane "had to circle to avoid the worst tropical storm I have ever seen. To the east, literally as far as one could see north and south, there was an advancing wall of inky cloud, a gray smoky curtain of water falling, and almost continuous lightning. We banked, swirled, dropped, bounced, and did everything but pick leaves off the trees — they were close enough below us, heaven knows. After we were in sight of Irumu, I noticed the co-pilot wiping his hands on the seat cushions. He had been doing a little sweating, too!"

And my own great rains? They were not as dramatic. I thought of a wet night in Amsterdam and going back to a tiny hotel that was like a ship, with canted floors and cabins for rooms. In Taipei, I waited for an hour under a shop awning while a warm rain fell. In an Indiana storm, I climbed into the attic of my old house to stop leaks and heard the voices of my wife and children rising like a benediction. So many rains, each beautiful and specific, and such a span of time. I heard a child ask what day it was and then exclaim, "Monday?

Again?" as if to say, "Does this just go on and on?" And I wrote:

A day of broken weather,
dead leaves
darkened with rain,
a sunlit interval
of sky streaked
by higher clouds . . .

Why should it seem
that something breaks
like thin ice in the heart?

It doesn't go on and on. But how can we ever leave the beautiful specificity of the world, this intricate home? I had not intended to write a farewell, but that note was always there. Bill wrote precisely, lovingly of the Congo, then left his letters behind. Mary, my wife's mother, grasped both our hands and said, "How will I ever get along without you two?" Then she went under deep sedation and left us. I wrote again:

. . . even the new tree
in its circle
of bricks and earth,

its fine lace glistening

with rain, was turning away,

but with such generous

and tender transience

I could let it go.

I saw also that I was making what poets call a "composition on a field," which I take to mean the discovery and mapping of many points until a new landscape arises dripping from the mist. And I came to wonder if we are not all charged, in some way, with the discovery and completion of the world.

Now the winter rains have come. "The harvest is past, the summer is ended," the prophet said . . . and I am content with that.

* * *

There was more I could have written about rain. How it has little taste but a freshness on the tongue. How it was here before us and will be here after us — weaving the seas, wearing down the mountains — from the improbable beginning to the unimaginable end. I could have written about the precise composition of amniotic fluid and of tears. There are philosophies in the rain, and nursery rhymes: "If all the trees were bread and cheese, what should we do for drink?" But there are no secret messages in it. Whatever we ask of it, it gives us that. Any other answers are written somewhere else, in a different hand. The rain is just the rain.

NOTES & ACKNOWLEDGMENTS

These notes are not exhaustive and are not footnotes to the text. They simply record some of the major sources of information. In most cases, only the basic internet URLs are given. The notes also serve as an acknowledgment of the very great help received from many people, during a project that depended heavily on e-mails and the internet. Thanks go especially to Nancy L. Bridges, for her fine rain and water photographs, to Lindsay Hadley and Tim Lisko for their handsome cover, and to Susanna Lippóczy Rich, who kept saying, "Tell me more." This book is dedicated to her.

CHAPTER 1 (Ancient Rain)

Crowley, T. J., & North, G. R. (1991). *Paleoclimatology,* Oxford.
Erickson, J. (1989). *The Coevolution of the Planet and Life.* Tab Books.
Guinness Book of World Records, 2005
World Book Encyclopedia
Froude, D., et al. (1983, Aug. 18). Ion microprobe identification of 4,100-4,200-myr-old terrestrial zircons. *Nature,* 304, 616-618.
Holliday, A.N. (2001, Jan. 11). In the beginning . . . . *Nature,* 409, 144-145.
Matsui, T. & Abe, Y. (1986, Aug. 7). Impact-induced atmospheres and oceans on Earth and Venus. *Nature,* 322, 526-528.
Mojzsis, S.J., et al. (2001, Jan. 11). Oxygen-isotope evidence from ancient zircons for liquid water at the Earth's surface 4,300 myr ago. *Nature,* 409, 178-180.
Valley, J.W., Peck, W.H. & King, E.M. (2002, April). A cool early earth. *Geology, 351-354.*
Watson, E.B., & Harrison, T.M. (2005, May 6). Zircon thermometer reveals minimum melting conditions on earliest Earth. *Science,* 308, 841-844.
Wilde, S. A., et al. (2001, Jan. 11). Evidence from detrital zircons for the existence of continental crush and oceans on the Earth 4.4 gyr ago. *Nature,* 409, 175-178.
E-mails: Dr. J. Thomas Howald, Franklin College, Franklin, Ind.; Dr. Gerald North, Texas A&M University; Dr. Simon Wilde, Curtain University of Technology, Perth, Australia.
Websites: www.pbs.org; www.cotf.edu; www.nsf.gov; science.nasa.gov; www.ozh2o.com; www.news.wisc.edu; www.solstation.com

A FINE SMIRR OF RAIN

CHAPTER 2 (The Shape of a Raindrop)

Bentley, W.A. (1931). *Snow Crystals*. McGraw, Hill.
Blanchard, D.C. (1998). *The Snowflake Man: A Biography of Wilson A. Bentley*. McDonald & Woodward.
Blanchard, D.C. (1966). *From Raindrops to Volcanoes*. Doubleday.
Eastaway, R. et al. *Why Do Buses Come in Threes: Hidden Mathematics of Everyday Life*. John Wiley and Sons.
Lynch, J. (2002). *The Weather*. Firefly Books.
Martin, J.B. (1998). *Snowflake Bentley*. Houghton-Mifflin.
Middleton, W.E.K. (1966). *A History of the Theories of Rain*. Franklin Watts; (1964). *The History of the Barometer*. Johns Hopkins.
Morton, O. (2003). *Mapping Mars: Science, Imagination, and the Birth of a World*. Picadore.
Newsom, H.E., & Jones, J.H. (Eds.). (1990). *Origin of the Earth*. Oxford.
Schaefer, V.J., & Day, J. A. (1981). *A Field Guide to the Atmosphere*, Petersen Field Guide Series. Houghton-Mifflin.
Simons, P. (1996). *Weird Weather*. Little, Brown.
Marshall, J.S., & Palmer, W.M. (1948). The distribution of raindrops with size. *Journal of Meteorology*, 5, 165-166.
McDonald, J.E. (1954, February). The Shape of Raindrops. *Scientific American*, 64-68.
Uijlenhort, R., et al. (2003). The microphysical structure of extreme precipitation as inferred from ground-based raindrop spectra. *Journal of the Atmospheric Sciences,* 60, 1220-1238.
E-mail: Dr. Nick Steph, Franklin College, Franklin, Ind.
Websites:www.straightdope.com [Walk or run in rain?]; www.worldclimate.com

CHAPTER 3 (Remembering Clouds)

Aristophanes, *The Clouds*, text at www.classics.mit.edu
Day, J.A. (2002). *The Book of Clouds*. Silver Lining Books.
Dillard, A. (1999). *For the Time Being*. Random House.
Schaefer, V.J., & Day, J. A. (1981). *A Field Guide to the Atmosphere*, Petersen Field Guide Series. Houghton-Mifflin.
Whittier, J.G. (1866). *Snowbound*. From *Bartlett's Quotations*.
E-mail: Dr. J. Thomas Howald, Franklin College, Franklin, Ind.
Websites: www.cloudappreciationsociety.org; www.cloudman.com; www.meteo.helsinki.fi/~tpnousia/nlcgal/nlcgal.htm [pictures of noctilucent clouds]; www.ems.psu.edu ["Bad rain" site]; lasp.colorado.edu; www.phobialist.com

146

NOTES & ACKNOWLEDGMENTS

CHAPTER 4 (The Sound of Rain)

Earth Observatory News (2001, Aug. 24). TRMM spacecraft getting a boost to extend its watch on weather and climate; (2001, Nov. 26). TRMM continues to provide diverse insights into climate on fourth anniversary, both at http://earthobservatory.nasa.com.gov/Newsroom

Minnaert, M. (1933). On musical air-bubbles and the sounds of running water. *The London, Edinburgh and Dublin Philosophical Magazine and Journal of Science*, 7, 235-248.

Nystuen, J.A. (2000, June 14). Listening to raindrops: Using underwater microphones to measure ocean rainfall. Article with full-text link at http://earthobservatory.nasa.gov/Study/Rain/rain_2.html

E-mails: Dr. Clifford Cain, Franklin College, Franklin, Ind.; Dr. Jeffrey B. Halverson, Joint Center for Earth Systems Technology, University of Maryland; Jonathan at Murni's in Bali, Ubud, Bali; Cynthia M. O'Carroll, NASA/Goddard Flight Center; Owner at Raindrums.com; Robert Vickers, publicist, Jetset Records, New York.

Websites: www.balikingdom.com; www.raindrums.com; users.chariot.net.au; www.apollosaxes.com; http://earthobservatory.nasa.gov

Music: *Echoes of Nature: Thunderstorm* [CD]. (1993). Delta Music, Inc., Santa Monica, Calif.

CHAPTER 5 (The Smell of Rain)

Bear, I.J., & Thomas, R.G. (1964, March 7). Nature of argillaceous odour. *Nature*, 201, 993-995; (1965, Sept. 25) Petrichor and plant growth. *Nature*, 207, 1415-1416.

The chemistry of rain. Text at starryskies.com/~kmiles/din/5-96/rain.html

Websites: science.howstuffworks.com; www2.abc.net.au; en.wikipedia.org; www.worldwidewords.org; www.wordsmith.org; www.greenfairy.com; petrichor.diaryland.com; www.truthorfiction.com [Blessing story].

CHAPTER 6 (Wettest, Driest)

Das, P.K. (1972). *The Monsoons*. St. Martin's.

John, B.K. (2004) *Under a Cloud*. Penguin.

World Almanac

Anderson, J.W. (2005, May 20). [Dead] Sea in death throes as waters recede. *Washington Post* article in *Indianapolis Star.*

Article on desertification at pubs.usgs.gov/gip/deserts/desertification/

Desertification and the Sahel, article at www.mrdowling.com/ 611-deserts.html

Desertification as a threat to Sahel, article at

www.eden-foundation.org/project/desertif.html
Second-year progress report for NASA research grant NAG5-12890
    "Limits of Life in the Atacama"; "Second experiments in the robotic
    investigation of life in Atacama Desert of Chile. Texts at
    www.frc.ri.cmu.edu/atacama
Sobhan, Z. (2004, June 5). Under a cloud: Life in Cherrapunji [review].
    Daily Star of Dacca, Bangladesh. Text at www.thedailystar.net
Spotts, P.N. (2003) Sahel drought: New look at causes. *Christian Science
    Monitor*, text at www.csmonitor.com/2003/1120/
Watters, R. (2004, Aug. 8). Cloud's own country [review]. *Tribune of India.*
    Text at www.tribuneindia.com
Websites: [wetness] www.albumindia.com; tourism-of-india.com;
    www.tiffinbox.org [John critic]; www.extremescience.com;
    www.hiohwy.com; www.ncdc.noaa.gov; [dryness] www.ladatco.com;
    www.space.com; gosouthamerica.about.com;
    www.thefreedictionary.com; http://magma.nationalgeographic.com

CHAPTER 7 (Drinking the Rain)

Columbia Encyclopedia.
Dale, S. (2003). Collecting fog on El Tofo Mountain. Text at
    www.fogquest.org
Johnson, D. (2002). Thirst and the drinking pilot. *Soaring.* Text at
    amygdal.danlj.org/~danlj/Soaring/Thirst.SoaringMag.html
O'Neill, E. (1914). *Thirst* [one-act play].
Philbrick, N. (2000). *In the Heart of the Sea: The Tragedy of the Whaleship
    Essex.* Viking Penguin.
U.S. Army Survival Manual FM 21-76. Text at www.basegear.com/ch16.html
McGee, W.J. (1906) Desert thirst as disease. *Interstate Medical Journal*, 13.
    At www.ag.arizona.edu/AZWATER/awr/mayjune01/readings.html
Websites: www.newint.org; www.survivalig.com; www.bbc.co.uk;
    team.abnamro.com; www.newstarget.com; www.bbcfactual.co.uk;
    college.hmco.com; wittingshire.blogspot.com;
    www.msc-smc.ec.gc.ca

CHAPTER 8 (The Sky Is Falling)

Dante Alighieri (1472). *The Inferno.*
Encyclopaedia Britannica (1911).
Gosse, P.H. (1902). *Romance of Natural History*. New Amsterdam Book Co.
King James Bible.
Simons, P. (1996). *Weird Weather*. Little, Brown.

NOTES & ACKNOWLEDGMENTS

Thwaite, A. (2002). *Glimpses of the Wonderful: The Life of Philip Henry Gosse.* Faber and Faber.

Letter: Walters, J.E.O. [On "marching frogs"].

Websites: www.viewzone.com; http://en.wikipedia.org; www.greece.org; www.history.navy.mil; http://stronghold.heavengames.com; http://news.zdnet.com; www.wonder-okinawa.jp; www.forteantimes.com; www.newadvent.org; www.madsci.org; www.acmeme.org; www.gettysburg.edu/academics [Dorothea Tanning poem]; www.strangemag.com; www.pgw.com; http://paranoid.about.com/library/weekly/aa082602b.htm [falling cow].

CHAPTER 9 (Death by Umbrella)

Cerf, B. (1944). *Try and Stop Me.* Simon & Schuster.

Chambers, R.L. (1864) *Book of Days.* Vol. 1, 241-244. Text of article, Umbrella History, at www.backyardcity.com/Umbrellas-Umbrella-History.html

Deaths, *North American Review* (July, 1815) Vol. 1, No. 2, 296.

Sangster, W., *Umbrellas and Their History.* Text at www.gutenberg.org/dirs/etext04/mbrll10.txt

Encyclopaedia Britannica (1911).

Bartlett's Quotations.

Websites: www.chinavista.com; http://encyclopedia.jrank.org [Prina]; www.hobby-o.com [Prina]; http://inventors.about.com; www.gutenberg.org; www.lewis-clark.org; www.baltimore.org; www.ratical.com [TUM].

CHAPTER 10 (A Garden in the Rain)

Cowie, D. (2004, July 31). His backyard garden: It's as right as rain. Knight Ridder Newspapers, in *Indianapolis Star*, E9 [Dan Welker rain garden].

Damon, B. & Maver. A. Chinese living water garden, at www.wholeearthmag.com

Gardening with water quality in mind, at www.inter.net

Hols, M. Rainwater garden: More than just a pretty space, at www.mninter.net

Rain gardens: A household way to improve water quality in your community, at http://clean-water.uwex.edu

Websites: www.raingardennetwork.com; www.nwf.org; www.ci.maplewood.mn.us; www.keepersofthewaters.org

A FINE SMIRR OF RAIN

CHAPTER 11 (The Great Floods)

Flood Fight of '50, Offutt's Pictorial Service, Vincennes, Ind., undated.
World Almanac.
Articles and maps on Yangtse River system, at http://encarta.msn.com
Dealing with the deluge, at www.pbs.org
Great Yangtse flood engulfs Hankow, AP story in *New York Times* (1931, August 22). 1.
Laris, M. (1998, Aug. 17). Untamed waterways kill thousands yearly, *Washington Post.*
Mauelshagen, F. Disaster and political culture at the German North Sea coast (since 1500), paper at historical seminar, University of Zurich.
Reynolds, Damaris Peck. Lindbergh's stay in Nanking, September 1931, at www.charleslindbergh.com [Reynolds' father was U.S. consul].
E-mail: Steven Harnsberger, China Connection, San Anselmo, Calif.
Websites: www.answers.com; www.chinabusinessreview.com; www.cnn.com/SPECIAL/1999/china.50/asian.superpower/three.gorges; http://en.wikipedia.org/wiki/Burchardi_flood; www.schillerinstitute.org

CHAPTER 12 (Dancing with Rain)

Adams, D. (1984) *So Long and Thanks for All the Fish,* Pan.
Ingersoll, E. (1928). *Dragons and Dragon Lore,* Payson & Clark. Text available at www.sacred-texts.com
Cold wave more intense [Cold and rain damage Ozark crops]. *New York Times* (1910, April 18).
China rain-making creates a storm, BBC News (2004, July 14).
History of Hopi snake dance, at www.brownielocks.com/snakedance.html
Rainmaking link to killer floods [Lynmouth disaster], BBC News (2001, Aug. 30).
Whipple, D. Blue planet: The geopolitics of water, UPI Science News at www.rff.org
Websites: www.aaanativearts.com [Olmec rain gods]; www.aip.org/history/climate/RainMake.htm [Early climate modification schemes]; www.gatewaystobabylon.com [Hadad]; www.answers.com; www.curtis-collection.com/hopiraindance.html; www.kshs.org [Melbourne, the Rain Wizard]; www.kuroshin.org; www.metafilter.com/mefi/43941; www.reference.com/browse/wiki/List_of_deities; www.rbs2.com/w2.htm [Early cloud seeding]; www.sandiegohistory.org [Charles M. Hatfield]; http://starryskies.com [for 1972 Rapid City flood]; www.witchaven.co.nz; www.worldspirituality.org/hopi.snake.dance.htm

NOTES & ACKNOWLEDGMENTS

CHAPTER 13 (The Language of Rain)

Bartlett, J. (1894). *A New and Complete Concordance or Verbal Index to Words, Phrases, and Passages in the Dramatic Works of Shakespeare, with a Supplementary Concordance to the Poems.*
Bartlett's Quotations.
Bridges, W. Arthur, "More Peculiar Than We Think." Unpublished poem.
Collected works of Sean O'Faolain, Ernest Hemingway, Somerset Maugham, T.S. Eliot, William Bronk.
Dixon, S. *Outer Begonia.* Unpublished novel.
King James Bible.
Maltin, L. (1999). *Movie and Video Guide.* Penguin, for information about *Singin' in the Rain* (1952), *Rains of Ranchipur* (1954), *Rain Man* (1988), *The Rainmaker* (1997).
Strong, J. (1894) *Exhaustive Concordance of the Bible.* Abingdon.

CHAPTER 14 (Rainy-Day Reading)

Middleton, W.E. K. (1966). *A History of the Theory of Rain,* Franklin Watts; (1964). *The History of the Barometer,* Johns Hopkins.
Blanchard, D. C. (1966). *From Raindrops to Volcanoes,* Doubleday; (1998). *The Snowflake Man: A Biography of Wilson A. Bentley,* McDonald & Woodward.

CHAPTER 15 (Rainy Days and Mondays)

Solomon, A. (1991). *The Noonday Demon,* Scribner.
Rosenthal, N. E. (2006). *Winter Blues,* Guilford.
Albert, P., et al. (1991). Effect of daily variation in weather and sleep on seasonal affective disorder. *Psychiatry Research,* 36(2), 51-63.
Rain, rain, at www.hr.ucdavis.edu
Reidel, J. (2005, Oct. 25). Sad conversation. *The View,* University of Vermont.
Rosenthal, N. E. (1993, Dec. 8) Diagnosis and treatment of seasonal affective disorder. *Journal of the American Medical Association,* 270:22, 2717.
Sanders, Michelle. (2005, April 4). Compare & contrast: Rainy days vs. Mondays. *The Heights.* Boston College.
Shields, Brooke. *Down Came the Rain: My Journey Through Postpartum Depression.* Christa, 2005.
Interview: Sheron Miller, mental-health worker, Franklin, Ind.

E-mails: Norman E. Rosenthal, Georgetown University Medical School;
Jerry Miller, Franklin, Ind. [rain songs, movies]; Kim Nielsen, Tonder,
Denmark.

Music: *The Essential Waylon Jennings.* [CD] Movie-Mars; Blockpoint.

Websites: www.authorsden.com; www.bestdoctors.com;
www.blogcritics.org; www.clinical-depression.com;
www.mind.org.uk; www.mcmanweb.com;
http://reintonation.blogspot.com

CHAPTER 16 (The Webb Farm)

Conversations: Sue Webb, Grosse Pointe, Mich.; Kathryn Webb, Joe
DeHart, Pam Parker, Franklin, Ind.; Carole McDaniel, Shelbyville,
Ind.; Steve Polston, Indianapolis, Ind.

Other: Dedication program, Zachariah Webb Wetlands. (2005, July 26).

CHAPTER 17 (In the Rain Forest)

Encyclopaedia Britannica. (1911).

Oxford English Dictionary [for first use in English of "rain forest"].

Johnson, D. (1999). *The Amazon Rain Forest.* Lucent Books.

World Almanac

Newsletter: Bridges, W. Andrew, in *The New York Zoological Society's
Newsletter: Belgian Congo Expedition.* (1947). Nos. 2, 3, 4.

Private letters: William Andrew Bridges to Lynn Vandivier Bridges. (1946).

E-mails: Rhett Butler, Menlo Park, Calif.; Steve Johnson, head librarian,
Bronx Zoo.

Websites: www.damsreport.com; www.mongabay.com;
www.rainforestweb.org; www.srl.caltech.edu; www.wrm.org.uy

CHAPTER 18 (The Trouble with Venus)

Philander, S.G. (1998). *Is the Temperature Rising?* Princeton.

O'Harra, D. (2005, Oct. 24). Alaska's summer backs case for global
warming. *Anchorage Daily News,* reprinted in *Detroit Free Press.*

Ridenour, D., Global warming-hurricane link just hot air, at
www.nationalcenter.org

Websites: www.bradenton.com; www.exxonsecrets.org;
www.geo.unizh.ch/wgms [World Glacier Monitoring Service];
www.globalwarming.org; www.nrdc.org; www.spusa.org;
www.worldwatch.org; www.worldwildlife.org

NOTES & ACKNOWLEDGMENTS

CHAPTER 19 (The Ice Miners)

Dansgaard, W. (2005). *Frozen Annals: Greenland Ice Cap Research.* Narayana Press, Denmark. Available only on line at www.nbi.ku.dk/side60066.htm

Augustin, L., et al. (2004, June 10). Eight glacial cycles from an Antarctic ice core. *Nature*, 429: 623-628.

Chinese ice cores provide climate records of four ice ages (1992, Nov. 30); Researchers date Chinese ice cores to 500,000 years (1997, June 19); Researchers in Himalayas retrieve highest ice core ever (1997, Nov. 21); Oldest ice core from the tropics recovered, new ice age evidence (1998, Dec. 3). All from Ohio State University Research News at http://researchnews.osu.edu

European project for ice coring in Antarctica (EPICA), at www.awi-bremerhaven.de

Maslin, M.A. & Ridgwell, A.J. Mid-Pleistocene Revolution and the "Eccentricity Myth." Paper, at http://tracer.env.uea.ac.uk/e114/publications/manuscript_maslin_and_ridgwell.pdf

McManus, J. (2004, June 10). A great grand-daddy of ice cores. *Nature*, 429: 611-612.

Spahni, R. (2005, Nov. 25). Atmospheric methane and nitrous oxide of the late Pleistocene from Antarctic ice cores. *Science*: 310, 1317-1321.

Walker, G. (2004, June 10). Frozen time. *Nature*, 429: 596-597.

E-mails: Jerry McManus, Woods Hole Oceanographic Institution.

Websites: www.agu.org; www.antarcticaconnection.com; www.climate.unibe.ch; http://en.wikipedia.org; www.esf.org; http://geology.rutgers.edu/~jdwright/JDWeb/1999/JDWright_NUREG.pdf; www.qra.org.uk; www.geology.sdsu.edu/how_volcanoes_work/Santorini.html

CHAPTER 20 (Written in Water)

Apocrypha [Ecclesiasticus].

Bridges, W. Arthur "A day of broken weather" in *Weedpatch or Jericho?* (1988). Private press. "What Was Passing" in *The Landscape Deeper In* (2005). Virtual Bookworm.

Keates, J. (1994). *Stendahl.* Sinclair-Stevenson.

Morton, O. (2003). *Mapping Mars: Science, Imagination, and the Birth of a World.* Picador.

O'Connor, M. *Montesano.* [Unpublished novel].

Letters: William Andrew Bridges, Michael O'Connor, Port Townsend, Wash.; Tammy Peng, Fullerton, Calif.; Ann Wakeling, Aviemore, Scotland; Linda Yew, Kuala Lumpur, Malaysia.

*The type for this book was set and formatted
by the author in Book Antiqua for the text
and Times New Roman for the end notes.*

*The author is happy to correspond with other
rainiacs. His addresses are 920 Walnut St.,
Franklin, IN 46131, and
william.bridges@insightbb.com*

Printed in the United States
62153LVS00003B/226-324

9 781589 399419